This book appears in the series Business and the Environment – an innovative series of interdisciplinary titles for use in Business and Management courses at all levels, under the general editorship of Denis Smith of Liverpool Business School.

The series is based on the redefinition of the concept of 'business environment', moving from a purely economic orientation towards an ecology-based model. It is a response to increasing consumer pressure to incorporate environmental concepts in the business education curriculum; and is intended to meet the demands of the new environmentalism by providing business educators with a range of texts covering the vital issues of the 1990s and beyond.

Dr Denis Smith is Professor of Management and Director of Liverpool Business School at Liverpool John Moores University. Following a first degree in the environmental sciences, he studied for Masters degrees in pollution control and business administration before carrying out research for a PhD in technology policy. He has previously held positions at the University of Manchester, Nottingham Business School, Leicester Business School and the Open University. He is Visiting Professor of Management at Kobe Business School, Japan and Sheffield University.

His research interests are in the areas of corporate responsibility, strategic management and risk & crisis management. He is the Editor-in-Chief of *Business Strategy in the Environment* and on the editorial boards of *Industrial Crisis Quarterly*, *Disaster Management*, and *European Environment*.

BUSINESS AND THE ENVIRONMENT:

Implications of the New Environmentalism

Edited by
Denis Smith,
Liverpool Business School
Liverpool John Moores University

Paul Chapman Publishing Ltd
144 Liverpool Road
London
N1 1LA

British Library Cataloguing in Publication Data

Business and the Environment: Implications of the New Environmentalism
I. Smith, Denis
658.408

ISBN 1 85396 159 0

Typeset by Selwood Systems, Midsomer Norton
Printed by The Cromwell Press, Broughton Gifford, Melksham, Wiltshire SN12 8PH

A B C D E F G H – 9 8 7 6 5 4 3

CONTENTS

NOTES ON THE CONTRIBUTORS

Dr Denis Smith is Professor of Management and Director of Liverpool Business School, where he is also head of the University's Centre for Risk and Crisis Management. He is the co-editor of *Waste Location: Spatial Aspects of Waste Management, Hazards and Disposal* (1992) (Routledge) (with M. Clark and A. Blowers). His main research interests are in the areas of corporate responsibility, risk assessment and crisis management.

Dr Anthony McGrew is Senior Lecturer in Government, Faculty of Social Sciences at the Open University.

Dr Paul Shrivastava is Howard I Scott Professor of Management at Bucknell University, USA and is a specialist in strategic management.

Professor John Coyne is the Bass Professor of Business and Head of Leicester Business School. His main research interests are in industrial economics, management buy-outs and acquisitions.

Dave Owen is Senior Lecturer in Accounting at the University of Leeds. He is a Chartered Accountant and has previously held academic positions at the Universities of Salford, Manchester and Otago (New Zealand). He has published many articles on social accounting topics in a wide range of academic and professional journals and is co-author of the leading text *Corporate Social Reporting: Accounting and Accountability* (Prentice Hall, 1987).

Dr Neil Hawke is Professor of Law at Leicester Polytechnic, Director of the Environmental Law Research Unit and editor of the *Encyclopedia of Environmental Health Law and Practice* (Sweet and Maxwell).

Jo McCloskey is a Senior Lecturer in Corporate Responsibility and Marketing within the Centre for Risk and Crisis Management at the Liverpool Business

School. She has previously lectured in marketing at Leicester Business School and in colleges in Ireland and Africa.

Bob Graves is undergraduate Programme Manager in the Business School at Nottingham Polytechnic. He has many years' experience as a lecturer to undergraduate and postgraduate business students. His interests lie in broadening the application of marketing, and he has acted as marketing adviser to several professional firms and public sector bodies.

Dr Frank Fischer is Professor of Public Administration at Rutgers University.

Geoff Essery is NE Safety and Environment Manager with ICI Chemicals & Polymers Ltd. His interests include major accident avoidance and emergency planning both on-site and in transport.

Dr Steve Tombs is a Senior Lecturer in the Centre for Risk and Crisis Management at the Liverpool Business School.

Dr Kenneth Green is a Lecturer in the School of Management at UMIST where he carries out research in the fields of worker safety, innovation and product harm.

Dr Edward Yoxen is with the Centre for Exploitation of Science and Technology, Manchester Science Park.

ACKNOWLEDGEMENTS

The last ten years have seen the emergence of environmental issues on to the agendas of a number of organizations. The publics in many western democracies have successfully begun to impose their views on the actions of business with regard to environmental degradation. One of the principal problems in changing the operating strategies of business has been the attitudes and values of senior managers. There are a number of approaches that can be taken to shift the views of business and these include pressure from the state, a more proactive role by the industrial interests groups and a shift in the education and training of managers. The latter process, that of increasing the level of concern amongst existing and future generations of managers, lies at the heart of a business school's mission.

Unfortunately, too few of the business schools have been proactive in this respect largely because of the lack of expertise amongst teaching staff and a weakness in the research base. In an attempt to both develop interest in this theme amongst business educators and provide a useful collection of material for teaching, a series of conferences were organized during the period 1990–91. This book has its origins in those conferences which were held at Leicester Business School in conjunction with the Business Education Teachers' Association (BETA) and sponsored by ICI Polymers Ltd., the Geneva Association and the Leicestershire Co-operative Society. I am indebted to my colleague Bob Graves for his early involvement in the project and to BETA for allowing me the opportunity to establish what turned out to be a very successful series of meetings. I am also grateful to Shirley Dexter at Leicester for her patience in coping with my handwriting and administering the project in its early stages. Thanks also for the use of the printer – you will not have to fill the paper tray up quite so often. At Liverpool, Joan Meadows brought order to chaos, and managed to find things in my filing system. Profound thanks are also due to her. It will be much easier next time, Joan!

As with all edited collections the book is a tribute to the work of the various collaborators and to them all I extend my thanks. I am particularly grateful to

Frank Fischer and Paul Shrivastava for their continued support and encouragement with regard to this book and also the series initiative. I also owe a debt of thanks to a number of people who acted as referees for early drafts of the chapters. I would like to express my thanks to Derrick Ball, Steven Batstone, Dominic Elliott, Bob Graves, Alan Irwin, Sandra Lofthouse, Don Lloyd, Tony McGrew, Dean Patton, Chris Prince, Ashok Ranchhod, Greg Riddell, Chris Sipika, Anna Soulsby, John Thompson and Steve Tombs, all of whom made some useful comments on chapters at various stages during the book's production. I would also like to express my warmest thanks to John Coyne and Tim Wilson for their unfailing support during my time at Leicester Business School. Tim Wilson, as assistant director of the Polytechnic, gave me the initial freedom to pursue the background research and supported the conference initiative from the start. John Coyne, as head of the business school, deserves a special vote of thanks. It is not often that the head of a large business school takes so much of an interest in a colleague's research that he then proceeds to green the building as well as the curriculum. Leicester stands as a model of what can be done in this respect. Without the level of support and encouragement that I received the book's development would have been a more painful process – thanks John. Heartfelt thanks are also due to Marianne Lagrange, from Paul Chapman Publishing, for her vision in promoting the book and the accompanying series – I hope that the end result was worth the wait.

Finally, I am forever grateful to Janice for her patience, support and understanding of my relationship with the wordprocessor – finally the umbilical is severed ... until the next time! This book is dedicated to her with much love.

Denis Smith
Liverpool, June, 1992

ABBREVIATIONS

ALARA	as low as reasonably achievable
ALARP	as low as reasonably practicable
APELL	awareness and preparedness for emergencies at local level
ASSC	Accounting Standards Steering Committee
BAT	best available techniques
BATNEEC	best available techniques not entailing excessive cost
BETA	Business Education Teachers' Association
BOD	biochemical oxygen demand
BPEO	best practicable environmental option
BPM	best practicable means
BSI	British Standards Institute
CAP	Common Agricultural Policy
CEO	Chief Executive Officer
CIA	Chemical Industries Association
EA	environmental assessment
EPA	Environmental Protection Agency
EPM	Environmental Protection Management
IBE	Institute of Business Ethics
ICC	International Chamber of Commerce
IPC	integrated pollution control
ISRS	International Safety Rating System
MAb	monoclonal antibody
NIC	newly industrializing country
PPP	polluter pays principle
SHE	safety, health, environment
TNC	transnational corporation
TQM	Total Quality Management
UNEP	United Nations Environment Programme

To Janice, with much love

1

BUSINESS AND THE ENVIRONMENT: Towards a Paradigm Shift?

Denis Smith

Introduction

The 1980s saw the emergence of environmental issues on to both corporate and political agendas throughout the western nations. In the UK the government proposed new legislation in order to arrest our continued slide towards widespread environmental damage. At the international level there were also a series of treaties put into place in an attempt to control the problem of environmental degradation. The political and corporate development of environmental issues seems destined to remain a major agenda item during the 1990s. As corporate groups struggle to maintain their legitimacy in the face of more stringent public demands, there is a growing need for business educators to give the next generation of managers the skills to cope with the demands of this new environmentalism. The present curricula for most management courses give little time to the question of 'green issues' (Smith and Hart, 1991) and it comes as little surprise, therefore, to find that a number of managers find some difficulty in dealing with such problems. For many, the question is one of perspective: they argue that all human activities result in environmental degradation of some form and, consequently, society has to strike a balance between its material desires and the requirement for a cleaner environment. The convenience afforded by the motor vehicle, for example, has been a principal factor in the increase of NO_x pollution, the increase in tetraethyl and tetramethyl lead deposits and the problem of waste materials arising from the consumption of oils and component parts. Similarly, the pollution of rivers and watercourses by sewage, generated within urban conurbations, has long been a problem. The growth in CFC pollution, the presence of dioxins in the food chain and the increasing burden of hazardous waste deposits are all testimony to the effects of unfettered economic growth. The desires of the consumer society have been allowed to encourage the increase in levels of such pollutants within the context of relatively weak international regulatory regimes. Consequently, the notion of sustainable growth necessitates changes being made in the core beliefs of society as well as those of industry. In order to sustain ourselves as a 'post-industrial' society we have to recognize that only through a process of 'green' development will it be possible to arrest the decline in environmental quality at the local, regional and global levels.

Throughout the latter part of the 1980s business became more actively aware of the growth in public concern over environmental issues. This concern was heightened by two of the biggest industrial accidents of all time. The first of these involved the release of methyl-isocyanate from the Union Carbide plant at Bhopal, India, which killed some 3,000 people and committed many thousands more to progressive debilitation and premature death (Shrivastava, 1992). The accident sent waves of panic through the chemical industry, particularly in the USA, as it shattered the belief that such accidents were beyond the limits of credibility. Whilst the long-term impact of the accident in many western countries was somewhat muted through physical separation, it nevertheless pointed to the fallibility of expert judgements and undermined the power of the technocratic élite (Fischer, 1991). The second accident also had a major impact on societal views regarding the acceptability of its host activity. In 1986 the nuclear power plant at Chernobyl in the USSR came close to a core meltdown – the oft-quoted 'China Syndrome' – and resulted in the release of large quantities of radioactive material into the environment. This 'radioactive cloud' drifted over most of western Europe, caused some 10,000 'immediate' deaths[1] in the USSR and may have contributed significantly to many more early deaths by cancer across Europe. In addition to such acute systems failures, there was also considerable concern expressed over so-called 'leper ships' which epitomized the problems associated with the international trade in hazardous waste. The problem stems from the potentially exploitable nature of poorly regulated industrial activity, as wastes from western countries were dumped in the developing world. The range and scale of incidents such as these undermined public confidence in certain sectors of industry and combined with concern over low-level pollution and product quality to put industry under severe pressure to literally 'clean up its act'.

One of the main problems associated with the whole issue of environmental pollution relates to source determination and the associated evaluation of causality. There is a view expressed within certain sectors of industry that the public does not understand the source of much of our pollution and that in relative terms the impact of industrial production on the environment is not as great as often perceived (see Table 1.1). However, the data in this table is far from complete and fails to detail a range of heavy metals, chlorinated hydrocarbons and other 'specific' substances and consequently the data needs to be treated accordingly. For substances such as SO_2 we find that power stations contribute the bulk of the problem with road transport being the prime source of CO pollution. In addition, it should be remembered that certain pollutants have a greater toxicity and longevity than others and many arise at point sources rather than from the more diffuse sources that are associated with urban areas. It is the large polluting point sources that often capture public attention and concerns because the damage caused is often 'easier' to detect.

Table 1.1 Pollution sources and composition (selected) in the UK

Source/ pollutant	Sulphur Dioxide	Nitrogen Oxides	Carbon Dioxide	Carbon Monoxide	Methane and Volatile Organic Compounds	Smoke
Power stations	71%	32%	33%	1%	–	5%
Industry	17%	13%	24%	6%	37%	17%
Agriculture	–	–	–	–	18%	–
Road transport	1%	45%	18%	85%	9%	34%
Domestic	4%	3%	15%	7%	–	42%
Commercial	3%	2%	6%	–	–	–
Refineries	3%	1%	3%	–	–	–

Sources: 1. Warren Spring/DTI data for 1988. 2. Howarth, J. H., derived from *Digest of Environmental Protection and Water Protection* (1989) no. 12, p. 30.

Given the complexity that surrounds environmental impacts, there is a need for business and public groups to engage in more open discourse about the nature of both acute and chronic pollution episodes. If industrial groups are concerned about public perceptions of their activities then they should be prepared to provide more information about their activities. In this context the UK badly needs a Freedom of Information Act that would remove the secrecy that has surrounded pollution regulation in the past. The difficulty in obtaining accurate information often fuels public concerns, and this is illustrated by the case of water quality data, with six years separating the release of government reports in 1985 and 1991.

In terms of pollution to watercourses, 23 per cent of all incidents arise from oil operations, 15 per cent from chemical (and other) industrial sources, with 17 per cent arising from farming and 19 per cent resulting from sewage discharges (*Digest of Environmental Protection*, 1989). In a report leaked to the UK media in October 1991, it was claimed that rivers in England and Wales had shown a considerable decrease in water quality since 1985 (nett 6 per cent with a maximum of 39 per cent in certain regions) (Ghazi, 1991). However, these statistics need to be qualified by noting that 11 per cent of the country's waterways, which were seen as being the most polluted in 1985, have shown an improvement in water quality. The implication here is that the largest point source polluters have improved their practices but that the overall trend is continuing downwards, with perhaps the greatest contribution to this trend arising because of government cuts in spending for sewage treatment programmes (Ghazi, 1991). The point that needs to be reinforced here is that pollution is a complex multi-faceted problem which covers a range of industrial and business activities, from agriculture to power generation, and is heavily influenced by societal and consumer demands.

Pollution as a Political Issue

Political and corporate concern for environmental quality is not a new concept. Indeed, some of the earliest concern can be traced back to the industrial revolution when a number of writers voiced an awareness of environmental degradation. Within the last forty years there have been numerous expressions of concern over the international dynamics of continued pollution. Whilst many of these objections have been likened to Malthusian fears over population growth, their significance has taken on a new dynamic with the scientific controversy over global warming and ozone depletion. The concerns expressed during the 1960s, represented through the publication of *Limits to Growth* (Meadows *et al.*, 1969), culminated in the now seminal conference on the environment held in Stockholm in 1972. For many, the Stockholm conference established the legitimacy of environmentalism as a political movement. Despite this early concern about the environment, however, the issue slipped down the political agenda during the 1980s only to re-emerge during the 1990s through the Bergen Conference (1990), the Second World Industry and Environmental Management Conference in Rotterdam (1991) and the 1992 UN Conference on Environment and Development. The obvious question that one needs to address at this juncture is 'Why, after twenty years, has the environmental question not been fully addressed?' The answer lies, in part, in the cyclical nature of policy issues (see, for example, Blowers, 1984). The oil crises of 1974 and 1979 combined with the onset of a world recession to submerge environmental issues within the international political agenda. It is only with the recognition of the scale, complexity and severity of the issues that public environmental concerns have again emerged as a potent political force (McGrew, 1990). Within this context there has also been a major shift in the driving forces for greater environmental safeguards. Whilst industry has long been seen as the root cause by many environmental groups, there is now a growing recognition that it also has the resources and expertise necessary to solve many of the environmental problems that currently beset our society. The issue has now emerged on to the policy agenda at public, governmental and corporate levels and there is a need for a partnership approach between these groups in order to ensure that the best policy options are followed to ensure environmental improvements.

The treatment of the environment within the academic literature has, in part, reflected the wider subjugation of environmental issues in world politics. Political scientists and international relations academics, for example, were preoccupied with super-power rivalry, Third World issues and domestic political upheaval throughout the 1970s and mid-1980s. The visibility and urgency of these issues served to overshadow the environment as a topic for academic discourse. Consequently, some writers have suggested that the issue has lain dormant at both the local and global levels (Blowers, 1984) whilst others have suggested that the complexity and multidisciplinary nature of the problems have conspired to encourage their relative neglect by some academic disciplines (see, for example, Simmonds, 1990). This is not to say that the issues have not been addressed and a quick glance at the literature would suggest that the academic coverage of environmental impacts has been wide and varied. However, almost certainly the more obvious environmentally concerned disciplines, such as geography and business studies, have been notable by their lack of fully fledged involvement (Kates, 1987; Simmonds, 1990; Smith, 1990). The academic business community in particular has, until very recently, been loath to take the matter seriously. Such neglect of an issue which is central to business is surprising at one level and yet understandable at another. Environmental concerns, like the vexed issue of corporate responsibility, run counter to the dominant, finance-based, business paradigm. As a consequence, we should not be too surprised that business practitioners and academics have shown a certain reluctance to fully address the challenges that these issues present. However, the corollary to this view is that the relationship between business and the environment lies at the heart of the wider environmental problem and that a feeling of being uncomfortable with the issues should not be a sufficient justification for their neglect. In its broadest sense, it is 'business', through the extraction, production and consumption process, which is responsible for a considerable amount of environmental degradation and any attempts at changing the behaviour of corporations will require a fundamental shift in the values and behaviour of managers. However, the caveat here is that society must also change its buying behaviour to accommodate the demands of the new environmentalism. It would seem churlish of society to demand that industry cleans up its act without making corresponding shifts in its own behaviour. Whilst the impact of individual behavioural change may seem trivial compared to that of a corporate body, the collective impact of such changes can be considerable in terms of environmental quality. The success of the Body Shop has illustrated that there is a demand from society for environmentally friendly products and, if the constituency of the green consumer can be widened, then more companies will see the market opportunities of being green and will shift the focus of their manufacturing accordingly. However, in the short term, by far the greatest improvements in environmental quality can undoubtedly be achieved by changing the behaviour of those corporations who are large point-source polluters.

There are two issues here. The first is that we must influence the current generation of managers who are controlling the business enterprise. If we are to achieve short-term improvements in the environmental performance of business we must encourage managers to recognize the immediate impacts that their actions can have on environmental quality. However, long-term improvements can only be made if we can change the way in which future generations of managers are educated. Such a strategic view of the problem will entail making major changes to the business curriculum at all levels and involve the retraining of existing business educators who have generally been over-exposed to the dominant financial paradigm. It was as a result of such attempts to 'green the business curriculum' that the idea for this book was born. Some of the chapters presented here saw life as presentations at conferences held in 1990, under the auspices of the Business Education Teachers' Association (BETA) in the UK. However, the book is more than a collection of essays and there are a number of connected themes which run through the various chapters and these need to be explored further.

Business and the Environment: Structuring the Issues

One of the difficulties in addressing environmental issues within business relates to the role of corporate culture and managerial values in affecting the corporate response. Mitroff and colleagues (Pauchant and Mitroff, 1988; Mitroff *et al.*, 1989; Pauchant, Mitroff and Lagadec, 1991) point to the difficulties involved in penetrating and understanding the relationship between corporate culture and the acceptability, or otherwise, of certain courses of action. They express the view that corporate culture is a multi-layered and highly interactive concept (Pauchant and Mitroff, 1988; Mitroff *et al.*, 1989; Mitroff and Pauchant, 1990) and, whilst their concern was with the relationship between corporate culture and crisis management, the concepts serve equally well to illustrate the points made within this book on environmental problems. The relationship between business and the environment, like the issue of corporate culture, can be considered in terms of an 'onion model' (Pauchant *et al.*, 1990) and Figure 1.1 illustrates the nature of this relationship and, at a conceptual level, the structure of this book. Echoing the concepts of corporate culture outlined by Mitroff and colleagues, the chapters in this book can be seen to approach the environmental problem at a number of levels.

Perhaps the most obvious level at which to approach the problem is in terms of *corporate actions and plans* for environmental improvements as these are visible representations of the corporate response. This is the main theme of the chapters by Tombs, Essery, and Green and Yoxen. Each of these chapters approaches the practicalities of the 'greening' process by looking at corporate actions within two areas of manufacturing – chemicals production and biotechnology. These industrial sectors are of importance within the context of environmental degradation. The chemicals industry has for many years been regarded by many as a serious polluter at both the local and global scale. Not surprisingly, therefore, any successful attempts to clean up the sector's 'dirty person' image would be a major achievement. The chapters by Tombs and Essery offer different perspectives on the issue. For Essery, who details the efforts made by ICI Chemicals and Polymers Ltd. to reduce pollution from its plants and with its products, the scale of the problem prevents a speedy and simple solution. He rightly argues that the industry's long history of pollution and the very substances that it produces creates immense problems for chemical companies. That many such companies are now committed to sustained environmental improvements is an important and necessary first step in arresting our continued environmental degradation. By contrast, Tombs presents us with a much more critical view of industry's activities in this respect. Whilst acknowledging the efforts made by companies such as ICI, he argues that the current activities of many other companies fall short of what is required in order to sustain and eventually improve environmental standards. For Tombs, a cosmetic

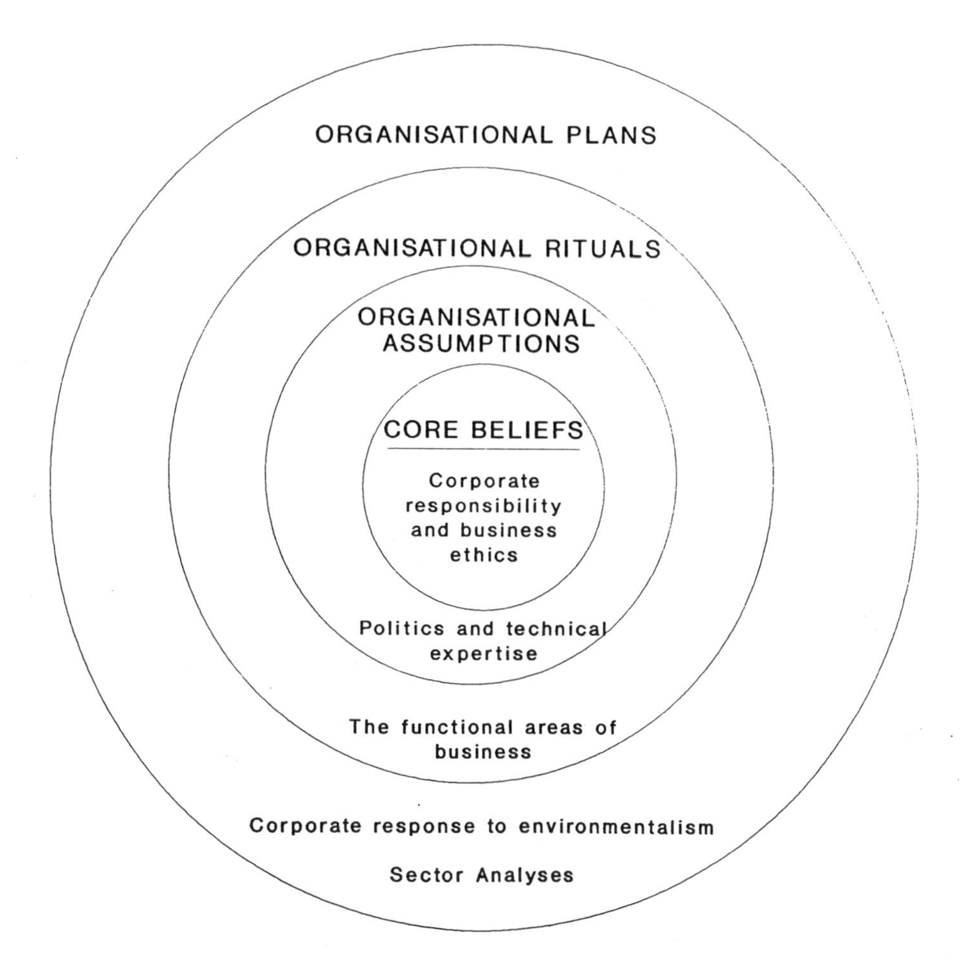

Figure 1.1 Business and the environment: structural dynamics of the issues.
Adapted from Pauchant, Mitroff and Ventolo (1990).

exercise, which is geared more towards public relations rather than real environmental improvements, is far from satisfactory. Instead, he offers an alternative and more radical strategy to that proposed by many in industry as a means of achieving sustained environmental performance. Finally, the chapter by Green and Yoxen points to the issues facing the biotechnology industry within the context of the new environmentalism. If chemical production has been a major polluter in the past then the biotechnology industry holds considerable *potential* to become a significant polluter of the future. The release of micro-organisms into the environment could have untold environmental consequences and this combines with the potential difficulties in developing effective regulatory enforcement to ensure that great care is taken to establish good environmental practices from the outset. The actions of industry in preventing pollution in this area of growth potential are important issues to address and Green and Yoxen set such an examination within a pan-European framework.

Moving beyond the practical manifestation of corporate environmental strategies, the book addresses the impact of environmental issues upon the functional areas of business. In particular, the book seeks to address four areas of academic study and application: economics, law, accounting and marketing. Economics and law are important issues because they represent the traditional areas of corporate conformity necessary to secure political legitimacy. If there are sound economic reasons for change or stringent legislative pressures are brought into force, then organizations will comply with societal desires for environmental improvements. Both of these pressures for change are generally seen as being external constraints on the organization in preventing managers from operating in a manner that would maximise short-term gain. Within the organization these external pressures for change become manifest in operating procedures and longer-term strategic plans. Both Coyne and Hawke point to important changes within the economic and legal regimes and assess their potential impact on organizations. In contrast the chapters by Owen and McCloskey *et al.* point to the accounting and marketing responses to such pressures. Accounting, for example, provides the mechanisms whereby companies can take the environment into their auditing procedures. Whilst accounting draws heavily on the underpinning economic principles and conventions, it nevertheless has a unique role to play within the greening process. Unfortunately the recession and continued requirements for growth during the 1980s have done little to encourage the incorporation of new economic thinking about the environment into accounting practice. The chapter by Owen addresses these issues and also points to the limitations that are inherent within current accounting procedures. By contrast, marketing, which represents the external image of corporate environmental policies, has been intimately involved in corporate responses to date. The scale of corporate claims concerning product performance regarding the environment is the main theme of the chapter by McCloskey *et al.* The current wave of environmental policies and societal demands have brought with them pressures for a more ethical marketing process and this issue is addressed within this chapter along with the effectiveness of basic marketing principles in dealing with environmental questions.

As we move to the deeper, more philosophical levels of the firm we are faced with addressing some of the key organizational assumptions, namely the issues of technical expertise and uncertainty within the context of environmental policy. Both Shrivastava and Fischer deal explicitly with these issues: Fischer advocating a more participatory form of expertise (especially in the context of environmental risks) and Shrivastava pointing to the need for a more ecologically-based definition of the business environment. Both of these issues combine with the political dynamics of the problem (as detailed by McGrew) to underpin the functional levels of business. The strategy employed in the past has been for industry to mount significant challenges to any threat to its legitimacy through the utilization of technical expertise or economic power (see, for example, Smith 1990a; 1991). However, within the current political climate, public groups are displaying an increasing intolerance of corporate strategies in this respect and have successfully mounted campaigns which seek to undermine the credibility of corporate expertise. Similarly the broadening of the corporate agenda to encompass more ecologically-based conceptualizations of the business environment should move the corporation into a more responsive culture which recognizes the validity of a wider stakeholder group. Such moves are, in turn, supported by the whole issue of corporate responsibility and business ethics which are dealt with by Smith. These issues lie at the heart of the organization and represent its core beliefs.

The issue of corporate legitimacy is held to operate at three levels, namely those of obligation, responsibility and responsiveness (Sethi, 1975). If the 1970s were dominated by issues of corporate obligation and the 1980s by calls for corporate responsibility then we can expect the next decade to deal with the more proactive issue of corporate responsiveness. Put simply, corporate bodies need to anticipate and prevent future societal concerns about their operations; in other words they will need to adopt a more strategic view of the problem. Such a paradigm shift in the culture of business will be difficult to achieve without the wholesale co-operation of managers, shareholders and business educators. A failure to incorporate a new set of environmental values at the heart of the corporate culture will result in a process of simply 'bolting on' a false consciousness in the form of a green tinge. This process will simply incubate the environmental crisis which will then re-emerge at a later date. A more fundamental inculcation of true environmental values within business is required and the chapters which follow address this issue in varying degrees of detail.

Towards Proaction

The aim of this book is to raise the profile of environmental concerns amongst the constituent groups involved in business education and practice. However, this chapter began by referring to the scale and complexity of environmental issues and raised the caveat that the complexity of the problem has hitherto served to hinder a thorough academic assessment of the issues. Consequently, this book can only serve to signpost issues for future debate and, hopefully, to stimulate those within the broad-based business community to re-evaluate their position with regard to the environment.

Inevitably, the book raises more questions than it is able to answer. The book has also omitted to deal with a number of areas explicitly within its coverage. Environmental auditing, for example, has been neglected here not because it is considered to be unimportant but rather because its sheer scale warrants a book in itself. Similarly, the human resource issue, whilst touched upon by Tombs in terms of worker safety, requires more detailed consideration elsewhere. In essence, the book represents the beginning of a discourse rather than the summation of it. It is hoped that the other volumes within the series, of which this is the first, will extend and further develop the concepts and issues raised here. The problem of environmental damage by business is a complex and multi-faceted issue which requires more than a technological fix to solve the problem. Whilst 'bolt-on, end-of-pipe' technologies will provide a brief respite from the immediacy of the problem, a more strategic move towards clean technologies and waste minimization is required for long-term improvements to be successful. This strategic approach will also necessitate a change in organizational culture where a quick-fix solution would be considered totally inappropriate. Long-term cultural change in industry will need to be underpinned by a similar shift in the approach taken by business education. For many years, the lack of suitable teaching material has proved an inhibiting factor in the development of a major shift away from the creed of profit over environment. This book, and the accompanying series, seeks to overcome this paucity of material – if it achieves that then it will have been successful.

Notes

1. Estimates suggest that in the period since the accident some 10,000 people have died from radiation-induced cancers. The final toll from the accident will not be known for many years due to the latency period for many cancers.

References

Blowers, A. (1984) *Something in the Air: Corporate Power and the Environment.* Harper and Row, London.

Digest of Environmental Protection and Water Protection (1989) no. 12, p. 30.

Fischer, F. (1991) Risk assessment and environmental crisis: toward an integration of science and participation, *Industrial Crisis Quarterly*, Vol. 5, no. 2, pp. 113–32.

Ghazi, P. (1991) Britain's rivers worse than ever, says watchdog, *The Observer*, Sunday 27th October, p. 11.

Kates, R. W. (1987) The human environment: the road not taken, the road still beckoning, *Annals of the Association of American Geographers*, Vol. 77, pp. 525–34.

McGrew, A. (1990) The political dynamics of the new environmentalism, *Industrial Crisis Quarterly*, Vol. 4, no. 4, pp. 291–305.

Meadows, H. D., Meadows, D. L., Randers, J. and Behrens, W. W. (1972) *Limits to Growth*, Earth Island Ltd., London.

Mitroff, I. I. and Pauchant, T. (1990) *We're so Big and Powerful Nothing Bad can Happen to us: An Investigation of America's Crisis-Prone Organizations.* Birch Lane Press, New York.

Mitroff, I. I., Pauchant, T., Finney, M. and Pearson, C. (1989) Do some organizations cause their own crises? The cultural profiles of crisis-prone vs. crisis-prepared organizations, *Industrial Crisis Quarterly*, Vol. 3, no. 4, pp. 269–83.

Pauchant, T. and Mitroff, I. I. (1988) Crisis prone versus crisis avoiding organizations, *Industrial Crisis Quarterly*, Vol. 2, pp. 53–63.

Pauchant, T., Mitroff, I. I. and Lagedec, P. (1991) Toward a systematic crisis management strategy: Learning from the best examples in the US, Canada and France, *Industrial Crisis Quarterly*, Vol. 5, no. 3, pp. 209–32.

Pauchant, T., Mitroff, I. I. and Ventolo, G. F. (1990) The ever-expanding scope of industrial crises: a systematic study of the Hindale Telecommunications outage, *Industrial Crisis Quarterly*, Vol. 4, no. 4.

Sethi, S. P. (1975) Dimensions of corporate social performance: an analytical framework, *California Management Review*, Vol. 27, no. 3, pp. 58–64.
Shrivastava, P. (1992) *Bhopal: Anatomy of a Crisis* (2nd edn.), Paul Chapman Publishing, London.
Simmonds, I. G. (1991) No rush to go green, *Area*, pp. 384–7.
Smith, D. (1990) Green into gold, *Times Higher Education Supplement*, 15th June, p. 14.
Smith, D. (1990a) Corporate power and the politics of uncertainty: risk management at the Canvey Island complex, *Industrial Crisis Quarterly*. Vol. 4, no. 1, pp. 1–26.
Smith, D. (1991) Beyond the boundary fence: Decision-making and chemical hazards, in Blunden, J. and Reddish, A. (eds) *Energy, Resources and the Environment*, Hodder and Stoughton, London, pp. 267-91
Smith, D. and Hart, D. (1991) The greening of business education in the UK. Paper presented at the 11th Annual International Conference of the Strategic Management Society, *The Greening of Strategy-Sustaining Performance*, Toronto, October 23–6.

2

THE POLITICAL DYNAMICS OF THE 'NEW' ENVIRONMENTALISM

Anthony McGrew

Introduction

The increasing salience of environmental issues on both the domestic and international political agendas is somewhat puzzling, coming as it does at the close of a decade which is more closely associated with the primacy of the market and consumerism than with the 'limits to growth'. Downs would argue that this is indicative of the fact that public concern for the environment largely reflects the business cycle, rising in prosperous times and receding when times are bad (Downs, 1972). This cyclical view is contested by Milbraith and Ingelhart, who identify in the contemporary political salience of environmental issues a dramatic shift (a 'silent revolution') in the core values held by political élites within all advanced industrial societies (Ingelhart, 1989; Milbraith, 1981). This chapter will explore the reasons behind the salience of environmental issues on the contemporary political agenda, investigating just how far it is valid to assert that advanced industrial societies are experiencing a 'new', more permanent wave of environmentalism. Such a task seems appropriate in the context of a book which is concerned with the implications of environmentalism for business and the context in which it has to operate.

The Emergence of 'Environmentalism'

Environmental issues, such as global warming, deforestation, disposal of toxic wastes, and acid rain to name but a few, have become regular features of contemporary media reports. Almost daily, governments and international agencies propose new initiatives to counter various threats to some aspect of the environment, from pollution to conservation. Moreover, when the final communiqué of the 1990 Western Summit meeting between the leaders of the world's seven major industrial states devotes considerable space to environmental issues the political primacy of the ecological crisis cannot be doubted. Yet this is in stark contrast to the political situation less than a quarter century ago when state inertia and institutionalized complacency prevented environmental issues from being seriously considered as significant public policy problems (Crenson, 1974). Even environmental pressure groups like Friends of the Earth in Britain or the Sierra Club in the United States were regarded as outside the normal political process, if not in some quarters an (insignificant) ideological challenge to the prevailing socio-economic order. Stereotypes abounded and the 'brown bread and sandals brigade' label served to marginalize environmentalists in the political arena. But today environmental issues have emerged as a 'new' dimension to the political agenda within all advanced industrial countries. What is more, political élites, traditional political parties and industrialists are embracing environmentalism with an enthusiasm which must surprise the most ardent environmental campaigners. Within the last year the British Conservative Party has been repackaged as the original party of 'conservation' and the Labour Party as a shade of Red-Green. Something significant underlies these changes: changes which are by no means confined to Britain but are products of a global process. Before investigating these underlying forces for change we need to explore the emergence of environmentalism as a distinct political ideology and social movement.

Environmentalism is not a new political cause; its roots reach deep into the nineteenth century and counter-reactions to the Industrial Revolution. Prior to the industrial revolution pollution was barely recognized as a political issue. Enloe notes that in the seventeenth century the Stuarts were petitioned about the environmental dangers to Londoners arising from coal burning in the capital, although legislative action was not taken until the enactment of the Clean Air Act almost two centuries later (Enloe, 1975, p. 23). In effect, it was not until the late nineteenth and early twentieth century that environmental issues really became politically salient.

It is possible to identify the origins of the modern environmental movements in both the United States and Britain in the counter-reactions to and critiques of early industrial capitalism. In Britain the great nineteenth-century thinkers Marx and J. S. Mill questioned the acceptability of unfettered growth and demonstrated an awareness of environmental issues (Walker, 1989; Hay, 1988). The cultural residue of the Romantic movement, which had championed a return to nature in the wake of the industrial revolution, contributed to a growing sensitivity to conservation, whilst some of the more popular writers of this period, such as Dickens and Mrs Gaskell, made much of the environmental degradation accompanying industrialism. But although there was a developing awareness of environmental problems this did not find expression on the political agenda or crystallize into an organized political form. Apart from the formation of the Smoke Abatement Society and some conservation groups which promoted a public awareness of environmental issues, there was little direct political activity connected with environmentalism. Interestingly this was not the case in the United States where, as Enloe relates, environmentalism was a powerful force at the turn of this century:

> American environmental politics has its roots in the conservation movement of the early twentieth century. Tied to the reformist Progressive movement's critique of capitalism's 'robber barons', fired by the end of the Western frontier, and led by such gifted politicians as Gifford Pinchot and Theodore Roosevelt, the conservation movement was one of the strongest political forces in the nation. But fifty years later, in 1954, the Sierra Club, direct descendant of the conservationists, had a mere 8,000 members.
>
> *(Enloe, 1975, p. 145)*

It was, however, not until the late 1960s that environmentalism became politically significant in most advanced industrial societies. This coincided with the prolonged period of post-war economic growth and the birth of vibrant counter cultures, of which environmentalism and feminism were but two. Such counter cultures challenged the fundamental structures and values of advanced capitalist societies. Environmental groups proliferated and environmentalism became a more self-conscious ideology which had at its core a revolutionary ethical philosophy: ecocentrism. As Hay defines it, ecocentrism articulates 'a belief that moral standing inheres in the non-human world, and that, conversely, the fate of other species is not to be arranged to suit the comfort and convenience of species homo sapiens' (*Hay, 1988*).

The revolutionary implications of this doctrine lie in its implied opposition to the exploitation of nature by humans. This carries with it a rejection of the dominant values of advanced capitalist and industrial societies: namely the striving for economic growth, materialism and human subjugation to technological systems. Modern environmentalism thus embraces a range of diverse causes which share a common concern for protecting the ecosystem from further degradation and safeguarding it for future generations. But environmentalism, as some believe, does not necessarily project a completely new or unique ideological alternative to established political ideologies on the left or right. On the contrary, ecological ideas have been incorporated into traditional socialist party programmes as well as Conservative and New Right programmes. Significant political ramifications for business flow from this since their natural political allies may no longer provide the degree of protection from environmental lobbies which, in the past, might have been expected. As Scott argues, it remains questionable whether ecological ideology actually provides a completely new synthesis which transcends the existing left-right ideological division (Scott, 1990).

During the 1970s environmental issues acquired a high visibility on the political agenda in Europe and the USA. This was partly a consequence of the success of environmental groups like Friends of the Earth, the Greens, Natural Resources Defence Council, the Ecology movement and others in mobilizing public opinion, particularly with respect to anxieties about nuclear power. Indeed the nuclear issue loomed large in the environmental campaigns of the period. But this was a period too of incipient political and economic crisis within the West which lent much credibility in the public imagination to the prognostications of environmental groups, such as the Club of Rome, of impending ecological crisis – a fear compounded by the events of 1973 and the dramatic realization of just how dependent industrial societies were on fossil fuel energy.

Besides mobilizing public opinion and shaping the political agenda, environmentalists also contributed to the institutionalization of environmentalism within the state apparatus. In both Britain and the United States, the 1970s witnessed the creation of the Department of the Environment and the Environmental Protection Agency respectively which remain responsible (but not exclusively so) for the formulation and implementation of policy on issues related to the environment. But, as Enloe indicates in her comparative study of environmental politics, there were other significant pressures behind the institutionalization of state regulation of the environment (Enloe, 1975). In particular the corporate world was quite responsive to the notion of national regulation. This was because industrialists recognized that environmental policy could not be formulated without their expertise and that a uniform set of national regulations would provide a 'level playing field' for all businesses as opposed to a proliferation of local controls.

The 1970s are often referred to as the decade of the environment. In terms of the political success of environmental movements in persuading governments in advanced industrial states to act on a range of environmental problems, particularly with respect to nuclear power, this may well be an appropriate label. But as the decade closed a new era of environmental politics commenced. In comparison with the early 1970s the early 1980s appeared to be a time of retrenchment since environmental causes figured less prominently on the political agenda. By 1977 public opinion findings had demonstrated a distinct decline in public concern with environmental issues (Dunlap and Van Leure, 1977). This is hardly surprising since inflation, economic recession and restructuring in most western economies, as well as the rise of a new market liberalism in the guise of the 'New Right', dominated the political scene. In Britain and the United States in particular, a new brand of conservatism posed a direct challenge to the environmental movement, since as Paehlke put it:

> neo-conservatism has declared [environmentalism] as an enemy of a kind with welfare bums and global communism. Environmentalism, like the welfare state and insufficient profit incentive is seen as an unnecessary and unwanted restraint on economic growth.
>
> *(quoted in Hay, 1988)*

Deregulation and a market philosophy, as well as an explicit commitment to economic growth, placed environmentalism on the political defensive more acutely than in the past.

One of the consequences of this changed political and economic context has been a restructuring and rejuvenation of the environmental movement not just on a national but on a virtually global scale. In Britain and the United States the environmental movement has become more highly organized and greater co-operation amongst major environmental pressure groups has emerged. There has also been a pronounced shift in the characteristic expression of environmental politics with a move away from citizen action groups towards much more institutionalized forms of political action such as pressure group activity, the contesting of local and national elections and the establishment of 'green' political parties. This deepening institutionalization of 'green' politics has also been accompanied by a significant shift in the philosophy and leadership style within the environmental movement. Hay notes that:

> Until the mid-70s the movement was dominated by scientific spokespersons who mounted a formidable challenge to the fundamental assumptions of industrial society on the basis of ecological science, but who, in innocence of the revolutionary social and political implications of such an analysis lamely advocated a politics of petition collecting and letter writing. But in the mid-70s scientific doomsdayism lost much of its steam. The focus shifted to questions of political theory and practice.
>
> *(Hay, 1988)*

The late 1970s and the early 1980s witnessed the culmination of these shifts in political strategy and organization. Perhaps the most convincing expression of this was the establishment in 1979 of the Green Party as a significant new force in West German politics.

Support for more traditional forms of political action has arisen from a new and growing political constituency in advanced industrial societies: what is often referred to as the 'new middle class'. This particular subset of the middle class is largely college-educated in professional occupations, relatively secure in material terms, and employed in public sector or more personal-service type organizations (Eckersley, 1989). The rejuvenation of environmental politics in the 1980s is, as will be discussed shortly, linked directly to the changing social structure of western societies. In turn these changes are connected with the emergence of a post-industrial order (Scott, 1990). This 'new middle class', born out of the transition to a post-industrial society, is the primary political constituency from which environmental movements now draw their most active support.

Alongside changes in support and organization the environmental movement has become increasingly internationalized in recent years (Porritt and Winner, 1988). This is part of a more general political strategy best captured in the phrase 'Think globally, act locally'. Very effective communication networks exist linking together environmental groups and parties across the globe. International co-operation, mutual support and assistance are now very much part of the politics of environmental issues. The most visible articulation of this international political co-operation has been evident in Europe. In a path-breaking development the various national Green parties have agreed and campaigned on a common European manifesto during the last two elections for the European Parliament. The intensification of this transnational activity by Green groups has also been evident in the mounting of international campaigns on environmental issues such as global warming and climatic change (Starke, 1990). Since most ecological issues are essentially transnational in character priority is now accorded to international political activity.

The recent rejuvenation of environmental politics can also be traced to the political impact of major environmental disasters. Chernobyl and Bhopal did much to reawaken public concern across the world to the environmental consequences of modern technologies. Moreover, growing awareness of the problems of industrial pollution, waste disposal, global warming, and the health risks associated with much of contemporary industrial activity has contributed to the deepening of environmental consciousness. Political action to protect or enhance the environment is now perceived to have greater legitimacy than at any time in the past, a fact which is evident in the intensification of domestic and international regulation of environmental matters.

As the 1980s proceeded environmental issues became increasingly politicized and internationalized. This trend is reflected most obviously in the intense international activity surrounding the long-term banning of CFC production induced by the perception of crisis associated with the problem of global warming – an issue which only five years ago was largely of scientific interest only. Such trends towards politicization and globalization combined with changing values amongst western publics and the strong reactions against the market-liberalism of the 1980s point towards more than just a simple rejuvenation of environmentalism in the 1990s. All the evidence suggests that we are experiencing not so much a revival of environmentalism as a dramatically new era of environmental politics, quite distinct from previous historical incarnations. But what are the manifestations of this 'new era' of environmentalism?

Manifestations of the 'New Era' of Environmentalism

This new era of environmentalism is characterized by a number of features which distinguish it from previous phases of environmental activism. Four features in particular stand out: public concern with environmental issues; the growth of green consumerism; the diffusion of ecological values; and the intensification of state regulation of environmental matters.

Public concern

As Table 2.1 indicates, a majority of the public in Europe consider environmental

Table 2.1 Attitudes of European public towards environmental protection

	Belgium	Italy	France	UK	Germany
Favourable	47.5	60.7	52.8	43.5	58.3
Mixed	29.4	27.6	33.3	38.2	31.3
Unfavourable	23.1	11.7	13.9	18.3	10.4
Total	100%	100%	100%	100%	100%
(n)	(830)	(934)	(836)	(1200)	(850)

Source: Rohrschneider, 1988

protection favourably. Moreover, what is significant about these figures is the intensity of concern expressed. For despite being asked to choose between economic growth and environmental protection, 'still a majority is evidently ready to bear the economic costs' (Rohrschneider, 1988). Furthermore, this deep concern for environmental questions shown by the general public is confirmed by other comparative studies (Milbraith, 1981).

Whilst survey evidence indicates that broad political support for environmental movements and political parties is strongest amongst the 'new middle class' groups identified earlier, it is not by any means confined to such groups (Cotgrove, 1982; Kreuzer, 1990; Rohrschneider, 1990). In Germany, for instance, Green voters were predominantly professionals, highly educated, aged on average 31 (Kreuzer, 1990). Yet, as Rohrschneider confirms, 'popular concern with environmental issues is by now widely spread and crosscuts most classes and cleavages' (Rohrschneider, 1990). Political support for Green parties comes from across the social spectrum but most particularly from those social groups outside 'the formal labour market or who have a flexible time budget such as the unemployed, housewives, university students, retired and marginally employed people and elements of the old ... middle class such as shop owners, farmers and artisans whose economic interests occasionally converge with those of the environmental lobbies' (Offe, quoted in Eckersley, 1989). However, in terms of environmental activists, as opposed to simply supporters, studies indicate almost conclusively that it is the 'new middle class' who tend to dominate environmental organizations. Environmental activism is thus drawing upon a particular social constituency within advanced industrial society, one which some studies suggest is likely to become increasingly significant in the transition towards post-industrial futures (Touraine, 1981; Offe, 1987; Scott, 1990). Public support for the environmental movement, as the evidence here suggests, is closely associated with the changing socio-economic configuration of industrial societies. It is therefore not so much a cyclical or temporary phenomenon but rather a consequence of more permanent socio-historical changes.

Green consumerism

Recent years have witnessed the emergence of green 'consumerism' and green 'productionism' as the public and business respond to growing environmental awareness. The Body Shop, Tesco, Sainsbury and other retail companies have their environmental consciousness as part of a broader corporate marketing strategy. In both the USA and Britain consumer organizations have adopted a 'green' approach. Ethical investment too provides a new approach to relating individual concerns to broader environmental concerns. Business has also responded to these emerging pressures, as can be seen in new marketing and product strategies. But it would be simplistic to suggest that such developments are at this stage little more than symbolic of the changing context of environmental politics within advanced industrial societies.

The diffusion of green values

In some senses 'we are all environmentalists now'. Eco-consciousness has penetrated and been assimilated by all the established political parties in Britain as well as in other western democracies. Furthermore, there is little evidence of any national political party in Europe or elsewhere seriously advocating unrestrained economic growth or deregulation of the environmental issue-area. Indeed, conventional political parties everywhere have adopted the policies of 'sustainable development' to advertise their ecological consciousness. These developments tend to confirm Scott's argument that new

> social movements effect change largely through influencing existing institutions of political intermediation, particularly policies parties ... In particular, by articulating new issues and by forming the new middle class into a political public, ecology has provided left-of-centre parties with an opportunity to modernize their political programmes.
>
> *(Scott, 1990, p. 152)*

If Scott is correct this is likely to have significant consequences for business. In particular, the incorporation of notions of 'sustainable development' into party programmes is likely to invite greater, rather than less, political regulation of business activities which have explicit environmental consequences. More generally 'sustainable development' implies some form of managed or acceptable growth strategies which are compatible with other political and social goals. The era of unrestrained economic growth, if it ever existed, is unlikely to survive into the twenty-first century at least amongst advanced industrial nations.

Increased state regulation

Despite the current trend in most advanced industrial states towards deregulation and limiting state intervention in the economy, a contrary trend is evident with respect to the environmental issue-area. Both the scope and intensity of regulatory activity designed to protect the environment appears to be on the increase. In Britain, where the move towards deregulation and self-regulation has been tenaciously implemented, recent studies seem to suggest a shift towards greater direct regulation of environmental matters (Baggott, 1989). Similarly in the United States local state governments during the Reagan years became increasingly active in establishing environmental legislation such that business has welcomed greater federal control as a means to impose some regulatory uniformity. This has also become increasingly the case in Europe. Moreover, the scope of international environmental 'legislation' has increased markedly over the last decade and is likely to increase further in the decade ahead (Newsom, 1989). As Mische confirms:

> Since 1921 more than 140 international treaties and other binding legal agreements related to the environment have been adopted ... All but two ... were adopted in the past 50 years, and more than half in the latter one-third of this fifty-year period,

indicating a rising trend in international agreements on environmental protection.
(Mische, 1989)

Alongside this global regulation, member states of the European Communities have been subject to an expanding array of regulations, directives and decisions covering environmental matters. Indeed, the Commission has identified the environment as one of the major policy areas in which more concerted community action and harmonization of national policies are required. A recent policy statement by the Commission stated that it 'will put strong pressure on the Member States to ensure community participation and coherence of national positions with Community objectives' (*Bulletin EC*, 1989). The consequence of this would appear to be greater regulation rather than less. But the implementation of environmental initiates remains a source of conflict and concern.

Taken together these four features – increased public concern, green consumerism, diffusion of green values, and increased state regulation – characterize a 'new era' of environmentalism, for they suggest that the environment is not an issue which simply reflects the dynamics of the business cycle but rather that it has its own momentum. A cursory examination of the underlying forces which condition the nature of contemporary environmentalism lends further support to this view.

The 'New Politics' of the Environment

How can we account for the distinctive character of contemporary environmental politics? What forces shape its dynamics? An examination of the current literature on environmentalism suggests that there are a number of interrelated dimensions which have to be taken into account in responding adequately to the above questions. These embrace: the existence of new social cleavages; the political strategies of the environmental movement; the contradictions of state policy; and the globalization of environmental issues. Each of these will now be elaborated briefly.

New social cleavages

Much existing literature considers the increasing salience of environmentalism to be a product of new social cleavages developing in industrial society (Touraine, 1981; Habermas, 1976, 1987; Offe, 1987). In simple language what is being suggested is that the social structure of industrial societies is being transformed as the nature of production and the economy itself changes. Post-fordist production techniques, flexible specialization and organization, and the shift to the service economy is restructuring, it is argued, the nature of industrial capitalist societies. This 'post-industrial' or 'dis-organized' capitalist society nurtures new social divisions, cleavages and interests. In particular the professional, administrative and technical intelligentsia – the 'new middle class' – has expanded rapidly. Since it is less closely enmeshed in industrial production than other social strata this 'new middle class' has embraced environmentalism as an expression of its own values and interests.

A rather different interpretation is placed on the origins of new social cleavages by Ingelhart who posits that the rejuvenation of environmentalism is the result of secular changes in the values held by western publics (Ingelhart, 1977, 1981, 1989; Ingelhart and Rabier, 1986). Western publics, he argues, are now less concerned with material goals and values than with post-material values which reflect concerns about the quality of life, including amongst other issues environmentalism (Ingelhart, 1981, 1989). This thesis is reinforced by extensive comparative survey evidence (Ingelhart, 1989). In many respects it reinforces the post-industrial society thesis although the engine of causation is somewhat different (Scott, 1990).

Whichever account of these new social cleavages and values is found most plausible the significant point is that both lead to the conclusion that greater public concern for the environment is not a temporary or cyclical phenomenon, but rather is connected to secular and structural changes in the nature of advanced industrial societies. In short, environmentalism is energized by the intersection of new social cleavages and shifting value systems. In this sense environmentalism represents a new social movement which is engaged in constructing a 'new politics' (Habermas, 1987). Recognition of this fact is central to understanding the political dynamics of modern environmentalism.

Political strategies of the environment movement

The environmental movement in Britain consists of a diverse and diffuse array of groups whose only common objective is a concern to protect the environment. Thus it includes Friends of the Earth, Greenpeace, the Green Party, the Conservation Society, as well as quangos such as the Nature Conservancy Council, the National Trust and others. Most of these organizations tend to be rather decentralized associations or networks and so lack financial resources and to a degree expertise. Unlike major producer and professional organizations, such as the Confederation of British Industry or the National Farmers Union, environmental groups have negligible power resources expressed in terms of financial, knowledge and expert capabilities. Nor is the British environmental movement atypical in any of these respects since as Porritt notes across the globe 'one would expect to find the same sort of diversity of groups and organizations as in the UK' (Porritt and Winner, 1988). It is this diversity and limited resource base which structures the political strategies adopted by environmental movements within the major industrial societies.

Environmental groups have only comparatively recently sought to engage in electoral politics at a national level. Despite the relative success of the various national Green parties in Western Europe contesting local, national and European elections, this remains a fairly 'novel' dimension of environmental politics. In general the political activities of the environmental movement are not dissimilar to those of other promotional pressure groups. Having neither the resources nor the collective organizational power of sectional groups, such as trade unions or business interests, they tend to effect political change by mobilizing publics through agenda-setting, politicizing issues, lobbying government and increasingly transnational pressure group activity. In addition they seek strategic alliances with other organized interests, such as trade unions or other promotional groups, in order to bring greater pressure to bear on legislators and government. A recent illustration of this can be found in the domestic debate in the USA concerning the proposed Free Trade Agreement with Mexico. Environmental groups, in both the USA and Mexico, joined forces and sought alliances with other organized interests opposing the agreement, such as farmers, unions, heavy industry, etc. Underlying this strategy was a concern that the more liberal environmental regulatory regime in Mexico would eventually dilute US regulations as companies relocated 'dirtier' activities. Clearly on this issue unions and business seeking protection from cheaper Mexican manufacturers and other exports had much in common with environmental groups in attempting to prevent full implementation of a free trade agreement. Recent agreement amongst the major industrialized states to act upon the problem of global warming reflects the culmination of a long campaign by a host of environmental groups, at both the national and international levels, to politicize the issue and mobilize opinion to such an extent that few governments could continue to ignore the pressure (Starke, 1990). In this political struggle to set the public agenda and mobilize societal pressure for governmental action the media, both domestic and foreign, is a critical channel of influence. This makes the new politics of environmentalism heavily reliant on the mass media, particularly the electronic media. Here is a major weakness since the media in most countries is reliant on business for advertising revenues. The overall structures of power may therefore militate against effective exploitation of media.

Contradictions of state policy

Although much of the activity of environmental groups such as Friends of the Earth and Greenpeace is directed at mobilizing public opinion, the ultimate aim is nearly always to instigate action by government whether local or national, or by international organizations. But one of the distinctive features of the role of the state in environmental politics is that there is a

> central paradox of an inherent, continuing potential for conflict between the state's role as developer and as a protector and steward of the natural environment on which its existence ultimately depends ... Posterity is a poor second to political survival or economic indicators.
>
> *(Walker, 1989)*

This central contradiction constrains the responsiveness of governments to the political demands of environmental lobbies and accounts for the fact that despite 'a growing literature on the steady-state society and its economic management, indiscriminate growth remains (un)official policy in nearly every nation' (Walker, 1989).

However, it would be erroneous to conclude from this that governments ignore demands and pressure for change since as has been noted already, this is patently untrue. Governments are constantly aware that state action or inaction has to be legitimized. One consequence of this is that they are forced to mediate between the conflicting pressures of economic as well as industrial requirements and the need to sustain their own political legitimacy. Whilst non-decision-making or crisis decision-making may be short-term strategies for dealing with these contradictory demands they are unlikely to provide any permanent resolution to what is a fundamental contradiction between the state's role in the economy and environmental matters. This contradiction is likely to become increasingly salient in shaping the dynamics of environmental politics since changing public attitudes will lead to greater demands on the state to protect its citizens from avoidable environmental hazards and degradation. This is not least because environmental pollution is 'no longer perceived merely as an aesthetic problem; it is now widely recognized as a serious health threat' (Eckersley, 1989).

On environmental matters, as on many other questions, the state should not be concerned as a monolithic entity. Rather it consists of different agencies pursuing their own bureaucratic interests. Thus there are many visible tensions and conflicts within the state apparatus with respect to the priority accorded to environmental matters and the substance of specific policy initiatives. Conflicts between those responsible for environmental health and those responsible for industry and the economy reflect different interests and policy goals within the state apparatus.

This matrix of contradictions and tensions which governments find themselves enmeshed in has important consequences for the character of environmental politics. For the state is a crucial site within which environmental politics is played out. The relationship between the state and environmentalism is thus extremely complex, with the consequence that there can no longer be an automatic assumption that economic logic ultimately determines the state's responses to environmental problems.

Globalization of environmental issues

Pollution recognizes no national frontiers. One of the consequences of this is that co-ordinated international action by states and international agencies is necessary to deal effectively with many contemporary environmental problems such as acid rain and climatic change. Over the last decade there has been an enormous increase in international political activity directed at environmental regulation and management (Soroos, 1986). The Stockholm Conference of 1972 left a legacy in the form of the United Nation's Environmental Programme which has acted as a catalyst for a burgeoning regime of international norms, rules and controls to protect the environment (Soroos, 1986). Environmental diplomacy now dominates the international agenda in a manner in which North-South issues did a decade earlier (Newsom, 1989).

What makes the current phase of environmentalism somewhat distinctive in a historical perspective is this intensification in the globalization of environmental issues and politics. It is also likely to become the critical dynamic in stimulating more extensive national regulation of environmental matters. This is for three reasons. Firstly, environmental groups have recognized the importance of exploiting their transnational networks and contacts to mount international campaigns on environmental issues (Starke, 1990). Thus in Europe, for instance, the Greens have co-operated together in pressuring the European Commission to formulate stringent European-wide environmental regulations. Most national Green movements by themselves would probably not have been able to pressure their own governments into similar action or else may have had to accept much less rigorous national legislation. It is thus hardly surprising that this transnational dimension to environmental politics is becoming increasingly salient in environmentalist political strategies. Secondly, business and industry too have recognized the opportunities which may arise from *international* as opposed to simply national regulation of the environment. In a study of the European chemical industry Grant notes that:

> The desire by West German public opinion for more stringent environmental regulation makes the German chemical industry very unhappy about the formulation of environmental policy in a national context. At the European level, the Greens are weaker, and other countervailing forces can be mobilized. At the very least, it can be ensured that the 'misery is shared' so that the German chemical industry is not disadvantaged by more stringent regulation than applies elsewhere in Europe.
>
> *(Grant, 1989)*

With the increasing internationalization of business the logic of more harmonized regional and global environmental regulation has a certain appeal since, as Grant suggests above, it prevents the costs being borne solely by one competitor in the global market. Alongside this, international trade by itself creates a strong sensitivity to foreign environmental regulations and pressures for greater harmonization. Thirdly, and finally, governments recognize that the absence of effective international rules governing environmental matters merely undermines national attempts to safeguard the welfare of their own citizens and the vitality of their national economies. Accordingly the tendencies are indisputably moving in the direction of more intensive and extensive international regulation. Environmental politics is becoming increasingly internationalized, if not globalized.

Conclusion

Just as environmental problems are unlikely to disappear from the political agenda in the near future so too is environmentalism unlikely to become marginalized in the political process. As this paper has sought to argue, a deepening environmental consciousness and the emergence of a 'new environmental politics' appear to be permanent features of advanced (if not post-) industrial societies. Moreover, as this 'new politics' increasingly permeates the language, institutions and structures of conventional politics in advanced industrial societies so too will it reinforce that on-going transformation of political culture which Ingelhart argues is an undeniable feature of these societies (Ingelhart, 1989). As Scott observes,

> It is the integration, and 'normalization', of previously excluded and 'exotic' issues such as ecology into mainstream politics that constitutes a fundamental shift in the character of conventional politics. Without this integration these issues would remain marginal.
>
> *(Scott, 1990, p. 151)*

That process of integration appears to be intensifying – a development which is likely to have profound implications for the business-government relationship in all advanced societies. Business is unlikely to escape much more intensive and extensive environmental regulation. Recognizing this, many of the major corporations within the advanced industrialized world have become increasingly active in the environmental policy domain. This is particularly the case for those operating in environmentally sensitive sectors, such as chemicals, manufacturing, etc. Moreover, growing economic interdependence increases the pressures on domestic business everywhere to conform to minimum environmental standards and to promote more uniform international environmental regulation so that, as noted earlier, there is some sense of a 'level playing field'. Given these pressures and, as this paper has argued, the existence of a more permanent environmental 'consciousness' within advanced industrial societies it would be somewhat curious if the 'environment', as a strategic issue-area, failed to figure prominently in the business studies curriculum of the 1990s.

References

Baggott, R. (1989) Regulatory reform in Britain: the changing face of self-regulation, *Public Administration*, Vol. 67, Winter, pp. 435–54.

Cotgrove, S. (1982) *Catastrophe or Cornucopia: The Environment, Politics and the Future*, Wiley, Chichester.

Crenson, M. A. (1974) *The Unpolitics of Pollution*, Johns Hopkins Press, Baltimore.

Downs, A. (1972) Up and down with ecology, *Public Interest*, no.28, Summer, pp. 38–50.

Dunlap, R. E. and Van Leure, K. D. (1977) Further evidence of declining public concern with environmental problems, *Western Sociological Review*, Vol. 8, pp. 109–12.

Eckersley, R. (1989) Green politics and the new middle class: selfishness or virtue? *Political Studies*, Vol. 37 pp. 205–23.

Enloe, C. H. (1975) *The Politics of Pollution in a Comparative Perspective*, Longman, London.

Grant, W. *et al.* (1989) Large firms as political actors, *West European Politics* Vol. 12, no. 2, pp. 72–90.

Habermas, J. (1976) *Legitimation Crisis*, Heinemann, London.

Habermas, J. (1987) *The Theory of Communicate Action*, Vol. 2, Polity Press, Cambridge.

Hay, P. R. (1988) Ecological values and western political traditions, *Politics*, Vol. 8, no. 2, pp. 22–9.
Ingelhart, R. (1971) The silent revolution in Europe, *American Political Science Review*, Vol. 65, pp. 991–1017.
Ingelhart, R. (1977) *The Silent Revolution*, Princeton University Press.
Ingelhart, R. (1981) Post-materialism in an environment of insecurity, *American Political Science Review*, Vol. 75, no. 4, pp. 880–900.
Ingelhart, R. (1989) *Culture Shift*, Princeton University Press.
Ingelhart, R. and Rabier, J. Rene (1986) Political realignment in advanced industrial society: from class based politics to quality of life politics, *Government and Opposition*, Vol. 21, no. 4, pp. 456–79.
Kreuzer, M. (1990) New politics: just post-materialist? The case of the Austrian and Swiss greens, *West European Politics*, Vol. 13, no. 1, pp. 12–29.
Mann, D. (1979) *Environmental Policy Formulation*, Lexington, Mass.
Milbraith, L. W. (1981) Environmental values and beliefs in the general public and leaders in the US, England and Germany, in D. Mann op. cit., pp. 43–62.
Mische, P. M. (1989) Ecological security and the need to reconceptualize sovereignty, *Alternatives*, Vol. 14, no. 4, pp. 389–427.
Morrison, D. E. and Dunlap, R. E. (1986) Environmentalism and élitism; a conceptual and empirical analysis, *Environmental Management*, Vol. 10, no. 5, p. 581–9.
Newsom, David (1989) The new diplomatic agenda, *International Affairs*, pp. 29–41.
Offe, C. (1987) Changing boundaries of institutional politics, in C. S. Maier (ed) *Changing Boundaries of the Political*, Cambridge University Press.
Porrit, J. and Winner, D. (1988) *The Coming of the Greens*, Collins, London.
Rohrschneider, R. (1988) Citizens' attitudes towards environmental issues, *Comparative Political Studies*, Vol. 21, no. 3 October, pp. 347–67.
Rohrschneider, R. (1990) New social movement: an empirical test of competing explanations, *American Journal of Political Science*, no. 1., February, pp. 1–30.
Scott, A. (1990) *Ideology and the New Social Movements*, Unwin Hyman, London.
Soroos, M. (1986) *Beyond Sovereignty*, University of South Carolina Press, Colombia.
Starke, L. (1990) *Signs of Hope*, Oxford University Press.
Touraine, A. (1981) *The Voice and the Eye: an Analysis of Social Movements*, Cambridge University Press.
Walker K. J. (1989) The state in environmental management, *Political Studies*, Vol. 38, pp. 25–38.

3

THE GREENING OF BUSINESS

Paul Shrivastava

> Avoiding environmental incidents remains the single greatest imperative facing industry today.
>
> *Edgar Woolard, CEO, DuPont*

> Environmentalism will be the next major political idea, just as conservatism and liberalism have been in the past.
>
> *Edith Weiner, Partner, Weiner, Edrich & Brown Consultants*

> The 1990s will be the decade of the environment.
>
> *President, Petroleum Marketers Association*

> Make environmental considerations and concerns part of any decision you make, right from the beginning. Don't think of it as something extra you throw in the pot.
>
> *Richard Clarke, CEO, Pacific Gas & Electric*

These are not the chief druids of Greenpeace talking. These are the voices of corporate leaders from *Fortune*, 12 February 1990 (Kirkpatrick, 1990). Over the past two decades the environmental movement has emerged as a powerful social, moral and political force with wide-ranging economic and organizational implications. Business organizations have been particularly affected by environmentalism. There is greater concern over conserving natural resources that form inputs into industrial production processes. Less environmentally polluting technologies and more environmentally safe or 'friendly' products are being demanded.

Doing business in an increasingly greening world poses new challenges for managers. These include anticipating demand for new environment-friendly products, designing safer, healthier and less polluting products and packages, developing less polluting manufacturing facilities, minimizing hazardous wastes, managing technological risks, conserving non-renewable natural resources, protecting the environment, and safeguarding worker and public health. How well are managers prepared for these challenges? Unfortunately, the answer to this question is negative. Managers in most companies are not trained to be sensitive to environmentalist demands. They deal with environmental issues grudgingly and with a compliance mentality. In fact, most managers prefer to let environmental, safety and health issues be handled by technical personnel at operating levels rather than devote top managerial attention and resources to them.

This lack of management preparedness and willingness to deal with environmental challenges is the focus of this chapter. It begins by reviewing the rise of environmentalism and its impacts on business, and then examines the anachronistic assumptions underlying current management practices on environmental problems. Finally, it discusses what it will take for management to respond adequately to the green challenge.

The Rise of Environmentalism

Concern for the environment has been heightened by widespread publicity given to impending ecological crises caused by acid rain, air pollution, the greenhouse effect, ozone depletion and hazardous wastes, and to major industrial accidents such as the Bhopal disaster, the Chernobyl nuclear accident, and the *Exxon Valdez* oil spill. Opinion polls in the USA, Europe and the Soviet Union show environmental protection as the primary political issue facing nations (Garelik, 1990; Gorbachev, 1990).

The rise of environmentalism has had major economic and political impacts. Its economic consequences include vast increases in expenditures to protect the environment. In the USA, private sector environmental protection expenditures have risen to over $200 billion in the past two decades. The Clean Air Act 1990 alone is estimated to increase pollution control costs by $20 to $50 billion per year. The government has raised budgeted expenditures for the US Environment Protection Agency (EPA) to an unprecedented $2.2 billion. An additional $8 billion is earmarked for the Superfund clean-up of hazardous wastes from 1987 to 1992. The 'environmental industry' that includes products and services for hazardous waste clean-up, environmental protection, industrial safety, environment-friendly products, etc. is expected to generate revenues in excess of $100 billion in this decade (Tolba, 1990).

Environmentalism congealed into a social movement in the 1960s. Today it has graduated into a potent political force. In the USA, one of President Bush's election platforms was to be the 'environmental President'. He signed the Clean Air Act 1990, which is the most comprehensive piece of environmental legislation in over a decade, and is elevating the EPA to the status of a Cabinet Department.

In Europe, Green political parties made impressive showings in national elections in the early and mid-1980s, particularly in West Germany, France, Norway and Sweden. By the end of the 1980s, most mainstream parties had added environmental protection to their own agendas. Norway's Prime Minister Mrs Gro Brundtland chaired the World Commission on Environment, which produced the influential report *Our Common Future*, a blueprint for environmentally sustainable economic development (McGrew, 1990). In the Soviet Union, President Gorbachev (1990) suggested that the threat from environmental annihilation is now greater than the threat of nuclear annihilation. He proposed wide-ranging domestic and international policy measures for environmental preservation.

Corporations in the USA have been forced by political and social pressures to deal increasingly with environmental protection, worker health and industrial safety issues. Pollutive and hazardous industries such as motor vehicles, chemicals, petroleum, nuclear power, hazardous waste management, etc. have seen a mass of new regulations. Motor manufacturers have increased their average miles per gallon to 27.5, from nearly half that in the early 1970s. They have added pollution control devices and passive seat constraints (belts, air bags) as standard features on cars. General Motors recently announced production of its electric car 'Impact' by 1993. The chemical industry has complied with new regulations to reduce environmental emissions from plants to less than 50 per cent of their 1970 levels (Shrivastava, 1991).

Despite these and many other changes, industry still falls short of societal expectations about its environmental performance. The public's expectations continue to rise, as do regulations and public pressures on corporations. If companies are to effectively meet societal demands in this area, managers will have to completely rethink environmental management needs.

Environmental Management Needs

The environmentalist impulse as perceived by corporations is not limited only to narrow demands for environmental protection. It broadly covers issues of industrial safety, environmental protection, natural resource conservation, protection of human health, and management of technological risks emanating from corporate activities. In response to these broad environmental concerns most large corporations today already deal with a wide range of tasks for which we will use the mnemonic SHE (Safety, Health, Environment). SHE activities are shown in Table 1.

It is clear from this list that SHE tasks cover a broad range of issues. This breadth is a positive feature of corporate environmental responses. What limits their corporate environmental effectiveness, however, is the way these tasks are implemented in practice. In most companies they are managed at operations level with relatively narrow technical focus, and generally ignored at the crucial strategic level of the firm.

SHE tasks must be dealt with both at operating and strategic levels within companies. At the operating level, management needs to develop operating policies and procedures, ensure adequate training and staffing of personnel, establish standards, monitor and reward performance. Also at operating level it must ensure compliance with relevant regulations and provide necessary information to government agencies and community organizations.

Table 3.1 Safety, health and environmental tasks

Industrial safety: Deals with making plants, warehouses and other industrial production systems safe for workers and the public. Involves choices of safety technologies, formulation and implementation of safety policies and practices.

Environmental protection: Deals with reduction of environmentally harmful emissions and effluents from industrial facilities. Also involves design and implementation of conservation measures that would lead to reduced use of non-renewable environmental resources, such as coal, oil, forests.

Waste management: Identification and remediation of hazardous waste sites. Programmes for minimizing waste generation.

Product integrity: Programmes for ensuring safety over the entire life of the product, from production and transportation to use and disposal of used product and its packaging.

Worker health: Policies and programmes aimed at ensuring healthy working conditions, compliance with occupational safety and health laws, monitoring health hazards emanating from products and production facilities.

Public and community relations: Communicating with external stakeholders and public on matters that affect company-public relationships. Co-ordinating emergency plans with community.

Media and government relations: Managing relationships with the media, regulatory agencies, and local, state and federal government.

Industrial security: Programmes for ensuring security of corporate assets. Loss-minimization programmes.

Risk, liability and insurance management: Programmes for minimizing technological and environmental risks. Minimizing corporate liability from product injuries, technological hazards, and hazardous waste sites. Insurance programmes to cover routine and extraordinary business risks.

Crisis management: Programmes for preventing incidents that could trigger corporate crises. Developing corporate capability for managing crises.

At the strategic level, management needs to develop SHE policies and an overall technology strategy. It needs to allocate resources, choose safety technologies, design technologically safe business portfolios, develop corporate-wide capability for dealing with crises, and ensure top management oversight for SHE-related performance.

These tasks now command the full-time attention of professional personnel, sometimes entire departments. Responsibility for these tasks now spans the entire corporate hierarchy sometimes reaching to top management levels. All these tasks have an impressive professional and technical knowledge base that needs to be mastered if managers are to deal with them proficiently. This knowledge is currently not part of formal business education programmes. Managers pick it up on the job, at professional meetings, and from industry colleagues.

Lack of knowledge is not the only limitation facing today's managers in performing SHE functions effectively. Equally important is the lack of ethical values and environmental sensibilities needed to make good managerial judgement and decisions on environmental issues.

If business is to deal with these tasks adequately, it will have to find ways of inculcating appropriate values and providing needed information to managers involved in environmental management.

Anachronistic Assumptions of Past Management Practices

Several fundamental assumptions that underlie corporate management of environmental issues, militate against effective management of environmental problems.

The business environment

The first assumption deals with the nature of the business 'environment'. In their decision-making and strategic frameworks few companies acknowledge the need to protect, enhance and renew the natural resources that are often so vital to their survival. Management theory and practice have adopted a peculiarly distorted definition of organizational environment. They erroneously view organizational environment as all economic, social, political, technological and commercial forces that influence the organizational performance (Fahey and Narayanan, 1984; Porter, 1980; Starbuck, 1976). This definition is biased towards economic performance and economistic thinking. It ignores the fact that 'nature' is the most fundamental environment of all human and consequently all organizational activities.

Elsewhere (Shrivastava, 1992), I have critiqued the existing concepts of organizational environment as being impotent and incapable of genuinely addressing the concerns of environmentalists. In a sense, the concept of organizational environment is CASTRATED by organizational researchers and managers. CASTRATED serves as a convenient mnemonic for critiquing assumptions underlying the concept of organizational environment. Here, I briefly review this critique to point out the need for fundamental reconceptualization of what managers think of as their firm's environment.

CASTRATED stands for Competition, Abstraction, Shallowness, Theoretical immaturity, Reification, Anthropocentrism, Time independent (ahistorical), Exploitable, and Denaturalized. The environment is seen as a legitimate venue or turf for '*competition*' among firms for resources. It implies antagonistic exploitative relations between organizations and their environments. It is viewed as an abstract social entity composed of *abstract* (and non-physical) economic, social, cultural and technological components. Natural, physical and concrete aspects of the environment are de-emphasized. The concept is *shallow* and *theoretically immature*. It is descriptively inadequate, lacks specificity and linkages to social and historical processes. Portraying the environment as a *time independent* variable, management theory and practice ignore the cumulative degradation of the natural environment over time. Management theory is premised on *anthropocentric* assumptions, and does not acknowledge the right of nature to exist, except for the well-being of humans. It *reifies* organizations and gives them higher status than that accorded to nature. It treats the environment as being *exploitable* without any limits.

This fundamentally flawed concept of organizational environment needs to be revised to allow managers and organizational researchers to incorporate environmentalists' concerns into managing organizations. As discussed later in this chapter, such a revised concept should be more biocentred and give centrality to nature and the planet's ecology as defining features of organizational environments (Pauchant and Fortier, 1990; Shrivastava, 1992).

The business organization as a beneficial system

The second assumption underlying management practice is that business organizations are generally beneficial, technological systems of production that serve the interests of many stakeholders. Consequently, all business activities are geared to improving productivity and efficiency, and benefits to stakeholders. This assumption ignores the basic fact that in trying to meet demands of multiple stakeholders, organizations are riddled with contradictions. They face contradictory pressures for profitability and safety, for long-term strategic investments and short-term financial performance, and for building complex, large and hazardous technological systems but minimizing risks.

As a consequence, organizational activities have many unintended destructive outcomes, such as environmental pollution, toxic wastes, hazardous products and work conditions, and technological risk to communities. These destructive effects have historically been treated as 'externalities' of production, to be dealt with by the state or public agencies. Corporations have thus avoided taking primary responsibility for managing the destructivity inherent in organizational activities.

In recent years this assumption has been challenged by the proliferating evidence of widespread environmental and public health destruction caused by corporate activities. Examples include depletion of stratospheric ozone by chlorofluorocarbons (CFCs), destruction of rain forests by unrestrained cutting of trees by timber and paper products industries, global warming caused by emission of chemical pollutants from power plants and motor vehicles, etc. Mounting evidence of environmental destruction forces us to acknowledge that business organizations are not simply systems of production, but are simultaneously systems of destruction. The destruction they cause includes both low-level chronic destruction of the natural environment, as well as sudden episodic large-scale environmental and health hazards such as the Bhopal disaster, the *Exxon Valdez* oil spill, and the Chernobyl nuclear accident (Likens, 1987; McKibben, 1989; Mitroff and Kilmann, 1984).

These negative outcomes of organizational activities would not have been a problem if the state/government was able to handle them effectively. However, in the past two decades, governments in most countries have been saddled with financial crises, and have had deficit budgets. Governments are simply not funded to deal with all the negative outcomes of private business activities. This realization has prompted a shift in the burden back to corporations, through environment, safety and health legislation.

Concept of risk

The third assumption underlying management practice involves the rather limited conception of 'risk'. Historically, firms have viewed risks primarily in *financial* terms. Financial risk refers to the probability of making financial returns on investments. Financial risk arises due to uncertainty about the future, volatility of financial markets, possibilities of inflation, imperfect knowledge of input-output relationships, etc. Organizations manage financial risks in relation to the size and nature of economic returns.

Firms are also concerned about *product market risks* that involve uncertainty about demand for their products. This type of risk is caused by changing economic conditions, consumer preferences, market demographics, competitive pressures, and regulatory changes.

This financial and product market conception of risks de-emphasizes the risks posed by technology, its location and its waste products. Technological and health risks emanating from business activities include risks of harm caused by industrial accidents, product injuries, occupational diseases, toxic wastes and environmental pollution. These hazards also impose risks of larger social, political and cultural disruption and upheaval.

Managerial discussions of risk predominantly examine risks *faced by* business organizations (and investors) and how to manage them. Little attention is paid to risks *posed by* businesses on their diverse stakeholders or the distribution of these risks among different sectors of society.

Just a technical problem?

A fourth limiting assumption of management is the inadequacy of technocratic orientation towards managing environmental problems. It is assumed that most of these problems are of a technical nature and are best handled by technical personnel at lower organizational levels (usually within plants). Hence, emphasis is placed on issues of technological design, safety and pollution control equipment, materials and supplies, safety and maintenance procedures, training of operators, etc. While this focus on technical elements is necessary, it is not sufficient for effective management of SHE problems.

In most corporations there is little (but growing) appreciation of the fact that SHE problems also have large social, political and emotional content. These elements of the problem require the attention of managerial personnel, and non-engineering solutions. They involve dealing with emotionally charged communities, conflict-ridden political systems, and poorly understood cultural and social norms. Narrow, well-structured and technical definitions of SHE problems may be easier to solve at lower operating levels, but only at the risk of ignoring these 'soft' issues. In very few companies do SHE issues receive the consideration of the Chief Executive, the Board of Directors or top management committees, where many of the soft issues are finally resolved.

Greening Business Management

To enhance the ability of companies to meet the environmental challenge, management needs to (a) reconsider its traditional values and assumptions, (b) develop new concepts for thinking about firm-environment relationships, and (c) organize environmental management functions with the same backing as they give to traditional functions of marketing, finance, production, human resources, etc.

Managerial values and assumptions about the environment

Many fundamental values about humans and their relationship to nature that we adhere to in our broader culture and in corporations are antithetical to the spirit of environmentalism. For example, the value of individualism stands in contrast to environmentalists' value of communitarianism (humans as part of a larger natural community), living in a harmonious coexistence with nature. Similarly, the cultural value of unrestrained consumerism contrasts with the moderated consumption-for-survival ethic of environmentalists. Anthropocentrism, a value that underlies virtually all human activities, is in conflict with the environmentalist value that accepts the absolute right of nature to exist and prosper independent of its benefits for humans (Carson, 1962; Devall and Sessions, 1985).

Table 3.2

Values about	Traditional	Environmentalist
Humans	Individualism Self-interest Independence Hierarchical	Part of a community Community interest Interdependence Web-like relationships
Nature	Inanimate External, separate from humans Exploitable	Living system Part of the human community Symbiotic
Relation of humans to nature	Anthropocentrism Subdue and conquer nature	Harmonious coexistence with nature Nurture and conserve nature

The differences between traditional values prevalent in organizations and the new environmental values that are emerging are depicted in Table 3.2. Differences between these values are a constant source of misunderstanding and conflicts between corporations and their environmentalist clients. People and organizations subscribing to environmental values have very different expectations from corporations.

To deal effectively with these expectations corporations need first to understand alternative value systems. These alternative values cannot be summarily rejected as radical or impractical. There are too many people (customers, suppliers, business associates) who now subscribe to environmental values for businesses to ignore them.

But understanding environmentalist values alone is not sufficient. Companies must make attempts to accommodate them by modifying their own value system and culture. This may require re-education of organizational members, and realignment of company products and technologies to more environmentally friendly alternatives (Pauchant and Mitroff, 1988).

A new concept of organizational environment

The limitations of received concepts of organizational environment were highlighted in an earlier section. If managers and organizational researchers are to seriously address the issue of the greening of business, they need a more nature-centred concept of organizational environment.

In a most basic sense, the environment connotes conditions that surround an entity. It is the locale or milieu for activities. Business organizations are economic institutions operating in a physical world. Hence, their relevant environment should be conceptualized as an economic biosphere, which includes not only economic, social and political influences but also biological and atmospheric ones.

Organizational environments include three elements: (a) the ecology of the planet Earth, (b) the world economic, social and political order, and (c) the immediate economic, technological and socio-political context of organizations.

Earth as the background of all human and consequently organizational activities is the most critical environmental element. The natural physical world including atmosphere, water, land terrain is the most fundamental base for all organizational activities (Brown *et al.*, 1987; Myers, 1984).

Human life in this physical world is structured into a world economic, social and political order established in the form of nation states with unique economic, social, cultural and political histories, and mutual relationships. Economic relations between nation states and within regional clusters of nations are governed by mutual treaties and international laws. Despite the apparent separateness of nation states, their economies are tightly linked and interdependent.

Within this world order operate organizations of various types pursuing diverse objectives. They are surrounded by economic, social, political, regulatory and technological conditions, opportunities and constraints. Historically, management has seen only these immediate surrounding conditions as *the* relevant 'environment' of the firm.

The expanded view of the environment advocated here also implies a new view of organization-environment relationships. Human activities including economic activities are a subsystem of the biosphere, rather than being separate and autonomous. The economy is inextricably integrated with the natural environment. Economic organizations as vehicles of economic activities are similarly integrated with their environments (Pearce, Markandya and Barbier, 1989).

Organization-environment relations do not involve passive reactive interdependencies, but rather two-way mutual influences that sustain each other. There is an embeddedness between the three environmental elements that makes them mutually interdependent. The influences between the organization and its environments are reciprocal. That means not only do these environmental elements influence organizational functioning, but organizations also have impacts on their environments (Shrivastava, 1992).

Environmental management capabilities

This reconception of organizational environment acknowledges the need to deal with SHE issues as being central and not peripheral to organizations. Accordingly, corporations need to develop expertise and capabilities in the following functions.

Corporate risk analysis and management

Many large corporations already have risk management departments in place. Most of these departments deal with financial risk through insurance coverage of traditionally defined hazards, such as natural disasters (earthquakes, floods, fires, hurricanes, etc.). This function needs to be organized around an expanded concept of risk which includes non-traditional sources of risk such as product injuries, technological hazards, political upheaval, environmental degradation and social changes (Kunreuther and Linerooth, 1983).

With the proliferation of complex new technologies and hazards, corporations today face significant technological and health risks which eventually translate into financial burdens. Managing these risks should not be limited to buying insurance. Indeed, insurance is a 'last resort' strategy, not one of primary containment. Unfortunately, many corporations, influenced by the 'polluter pays' principle, see financial insurance as being sufficient. This is a short-term view of the technological risk problem. A more far-reaching solution involves taking preventive actions which in turn implies a new management culture of safety. It should be expanded to include risk reduction through redesign of business portfolios, use of inherently safe technologies, design of environmentally safe products and technologies, creation of compensation systems for damages caused by corporate hazards, and the development of alternative conflict resolution mechanisms.

Crisis management

Corporate crises arise from many diverse sources, such as environmental pollution incidents, industrial accidents, product injuries, product sabotage, occupational diseases, hostile takeover, labour strike, market decline, foreign competition, etc. Managers have traditionally neglected developing crisis management capabilities, because the probability of these crises occurring is very low (Pauchant and Mitroff, 1988).

Traditional contingency planning is often not enough to deal with the mega-crises that can engulf modern corporations. There is a need for comprehensive crisis management systems, that proactively manage both pre-crisis and post-crisis phases. These include early warning systems, prevention planning, crisis and emergency management, and business recovery. Without such systems companies are unprepared to deal with even minor incidents, which can escalate into full-blown crises (Roberts, 1989; Smith, 1990).

Environmental policy and management

Most companies respond to environmental issues in the form of compliance to government regulations. In the past two decades 'environmental laws' have proliferated and so has the regulatory burden on companies. If companies continue to take an antagonistic position on regulations, they will continue to be burdened with ever increasing regulations.

In dealing with regulations, companies need not be reactive and compliance oriented. Instead they can be proactive and set voluntary standards of safety and environmental protection that reduce the chances of ill-conceived regulations. This requires understanding of the regulatory/public-policy-making process, and engaging it constructively and in a co-operative spirit to shape national environmental policies. An example of such participatory regulation is DuPont's involvement in the regulation of ozone-destroying chlorofluorocarbons (CFCs). As the largest producer of CFCs, DuPont has the technological know-how to create CFC substitutes, establish standards and co-ordinate an orderly transition to safer substances. It took a leadership role in developing national and international regulation of CFCs by sharing its expertise with governmental, international and community organizations.

Corporate and business strategies for environmental protection

Corporations can develop innovative approaches to preserving environmental resources and protecting the environment. These management strategies and techniques, such as Zero Pollution Production Strategies, Technology Portfolio Analysis, Emergency Planning, Environment, Safety and Health Auditing, Worst Case Scenario Analysis, Crisis Prevention Planning and Crisis Decision-Making Simulations, can reduce corporate exposure to risky technologies. The important point here is to integrate hazard avoidance and management in the strategic thinking of the company.

Public, media and government relations

Organizations must not only be responsive to the new SHE demands placed on them by society, they must also *be seen to be responsive.* This requires dissemination of information about organizational SHE activities to communities living around company facilities, consumers using company products, the public at large and government agencies in charge of monitoring them. The media act as an important conduit for this communication. Managing these communications requires establishment of a two-way communications programme through which the organization and its stakeholders can discuss mutually relevant risk issues.

Conclusion

This chapter represents a small first step in helping business to meet the challenges of environmentalism. Rather than providing conclusive answers, it opens up a host of new questions for inquiry. The new pressures on companies posed by the rise of environmentalism will require far-reaching changes in corporate philosophies, strategies, structures and systems. The magnitude of changes needed is implicit in the pronouncements by CEOs cited at the beginning of this chapter.

Environmentalism may turn out to be the next major organizing concept for society in the twenty-first century. In a post-industrial, post-modern, affluent world, environmentalism could be the substitute for ideals of social welfare and poverty eradication. Environmental annihilation could be a more real risk than nuclear annihilation. Environmental politics could dominate geopolitics. What will this imply for businesses, business studies and business educators?

At the risk of exaggeration let me suggest that the challenge facing business managers is of revolutionary proportions. It calls for wholesale rethinking and reformation of business values, objectives, strategies, products and programmes. Some initial steps towards this rethinking are presented in this chapter. Managers need to assess, discuss and implement these ideas. If they fail to do so they risk being left behind in a rapidly greening business world.

References

Brown, L., Chandler, W. U., Flavin, C., Jacobson, J., Polock, C., Postel, S., Starke, L., and Wolf, E. C. (1987) *State of the World 1987*, Worldwatch Institute, Washington D.C.

Carson, R. (1962) *Silent Spring*, Fawcett, Greenwich, Conn.

Devall, B. and Sessions, G. (1985) *Deep Ecology: Living as if Nature Mattered*, Peregrine Smith Books, Salt Lake City, Utah.

Fahey, L. and Narayanan, V. K. (1984) *Environmental Analysis*, West Publishers, St Paul, Minn.

Garelik, G. (1990) The Soviets clean up their act, *Time*, 29 January, p. 64.

Gorbachev, M. (1990) Address to the Global Forum on Environmental Protection and Development for Survival. Moscow, 20 January.

Kirkpatrick, David (1990) Environmentalism: the crusade, *Fortune*, 12 February, pp. 44–51.
Kunreuther, H. C. and Linnerooth, J. (1983) *Risk Analysis and Decision Processes*, Springer-Verlag, New York.
Likens, G. (1987) Chemical wastes in our atmosphere: an ecological crisis, *Industrial Crisis Quarterly*, Vol. 1, no. 4, pp. 13–33.
McGrew, T. (1990) The political dynamics of the new environmentalism, *Industrial Crisis Quarterly*, Vol. 4, no. 4, pp. 291–305.
McKibben, R. (1989) *The End of Nature*, Random House, New York.
Mitroff, I. I. and Kilmann, R. H. (1984) *Corporate Tragedies*, Praeger, New York.
Myers, N. (ed.) (1984) *Gaia: An Atlas of Planet Management*, Anchor Press, Doubleday and Co., New York.
Pauchant, T. and Fortier, J. (1990) Anthropocentric ethics in organizations, strategic management, and the environment, in P. Shrivastava and R. Lamb (eds.) *Advances in Strategic Management*, Vol. 6, JAI Press, Greenwich, Conn.
Pauchant, T. and Mitroff, I. I. (1988) Crisis prone versus crisis avoiding organizations: is your company's culture its own worst enemy in creating crises? *Industrial Crisis Quarterly*, Vol. 2, no. 1, pp. 53–64.
Pearce, D., Markandya, A. and Barbier, E. (1989) *Blueprint for a Green Economy*, Earthscan Publications, London.
Porter, M. E. (1980) *Competitive Strategy*, The Free Press, New York.
Roberts, K. H. (1989) New challenges to organizational research: high reliability organizations, *Industrial Crisis Quarterly*, Vol. 3, no. 2, pp. 111–27.
Shrivastava, P. (1987) *Bhopal: Anatomy of a Crisis*, 2nd ed., Paul Chapman, London.
Shrivastava, P. (1991) Corporate self-greenewal: strategic responses to environmentalism. Paper presented at the Academy of Management Annual Meeting, Miami, August.
Shrivastava, P. (1992) CASTRATED environment: greening organizational science, *Academy of Management Review* (forthcoming).
Siomkos, G. (1989) Managing product harm crises, *Industrial Crisis Quarterly*, Vol. 3, no. 1, pp. 41–60.
Smith, D. (1990) Beyond contingency planning: towards a model of crisis management, *Industrial Crisis Quarterly*, Vol. 4, no. 4, pp. 263–75.
Starbuck, W. H. (1976) Organizations and their environments, in M. D. Dunnette (ed.) *Handbook of Industrial and Organizational Psychology*, Rand McNally, Chicago, pp. 1069–123.
Tolba, M. (1990) Keynote speech to Globe 90. GLOBE'90, Vancouver, B. C., 19–23 April.
World Commission on Environment and Development (1987) *Our Common Future*, Oxford University Press, New York.

4

ECONOMIC CONCEPTS AND ENVIRONMENTAL CONCERNS: Issues Within the Greening of Business

John Coyne

Introduction

The increased attention given to environmental factors in society at large has led educationalists to a re-examination of the syllabuses for programmes of study in order to make them more relevant to the major issues of the day. This is particularly so in business education where the effects of the new environmentalism on consumers and producers has led to changes in the way in which certain subjects are approached and has even produced new sub-branches of disciplines. These changes have so far been modest in the UK with few major universities and polytechnics in 1990 having brought about major adaptations to syllabuses (Smith and Hart, 1991). Nevertheless, the debate has been engaged and the issues surrounding the form that adaptation should take have begun to be considered. It is important that educational programmes fully reflect the major concerns facing society and in the sphere of business there is little doubt that the 1990s is likely to be the 'E' decade – one in which the twin pillars of ethics and environment are at the forefront. How one changes business programmes to reflect these 'new' or 're-emerging' concerns is a major question.

It appears there are two basic approaches which can be taken in assessing the adaptive process. First one can simply add environmental concerns to the list of topics in any syllabus and 'bolt on' the issues which face that particular discipline. Alternatively you can step back from the accepted material of the discipline and ask a more fundamental question as to how the assumptions on which the discipline is based stand up to a concern which encompasses environmental issues. We may then assess whether there are fundamental weaknesses or simply a need for caution in the application of those disciplines. If the latter course of action is taken then there will be some disciplines which stand up better than others.

This chapter looks at some of the accepted elements of the neo-classical paradigm in economics and addresses the way in which a concern for the environment has been incorporated within it, or perhaps more accurately, at the way in which the paradigm has been used to address environmental issues (Pearce and Turner, 1990). It goes on to suggest that, with respect to the major questions facing the ecosystem as a whole, the restrictive assumptions on which the models are built have a tendency to leave some of the key questions dangerously indeterminate. Some may go so far as to say that these models neatly sidestep the major questions. There is no evidence to support the view that the paradigm is inapplicable to the problem or inaccessible to policy-makers, indeed certain elements are elegant in their simplicity, but we can question whether economic forces alone will ever be sufficient to explain, and therefore form the basis of, policies designed to ensure long-term environmental stability.

The Basic Economic Framework

In a chapter of this kind it is never going to be possible to review the whole of neo-classical economics and the way in which it has approached environmental issues. However, certain key elements on which the models are built will be examined to show the way in which they may be used to address environmental concerns and to identify any deficiencies these models may have in providing economics-based solutions for perceived global problems. It is necessary first to set out some goals for environmental concern, second to briefly collapse the neo-classical paradigm into a compressed, illustrative form, and finally to extract elements arising from this compression for examination and evaluation.

The Goals of Environmental Concern

When protection of the environment is considered in the context of the rising demands made upon it by a society which is growing in numbers and in consumption needs it is inevitable that one seeks to find a balance between ecology and development. The Brundtland Commission described our concern for the future as a concern for 'sustainability – the ability to maintain the desirable elements of the status quo into the future' (The Brundtland Report, 1987). Sustainable development was defined by Robert Allen (1980) as follows:

> Sustainable development – development that is likely to achieve lasting satisfaction of human needs and improvement of the quality of human life.

A society which exists within this sustainability has been defined by James Croomer (1979) as

> one that lives within the self-perpetuating limits of its environment. That society ... is not a 'no growth' society ... It is, rather, a society that recognizes the limits of growth ... [and is one which] looks for alternative ways of growing.

In these definitions one sees a balance between the interaction of society and the resources of the natural environment in such a way as to produce for society the optimum level of consumption which it requires whilst not depleting the stock of environmental goods for future generations. In this way inter-generational transfers are neutral and the development is sustainable. If we look at the background against which this idea must be considered it clarifies the magnitude of the task. The consumption needs of society depend upon two factors: the rate at which the consuming population is growing and the average propensity to consume of any given population at a point in time. To take the first factor, global population is currently growing at a rate of one million more births than deaths every four days. An editorial in *The Lancet* placed population growth in stark terms when it said, 'If a bomb as destructive as the one that destroyed Hiroshima had been dropped every day since 6 August 1945, it would not have stabilized human numbers' (*The Lancet*, 1990). The consequent demands placed by such population growth on sustainable development are severe indeed. If we then examine patterns of consumption we note increasing demands, aspirations and energies devoted to the acquisition of ever greater levels of material wealth. A standard of living enjoyed by perhaps one-fifth of the world's population is not unnaturally a level of aspiration for the remaining four-fifths. In such a context the average level of consumption is likely to rise considerably, even if we only allow for parity of consumption between nations. Thus on one side of our equation of sustainability there are massively rising demands by the consuming society for the transformation of resources into goods that satisfy current and prospective needs.

Against this scenario the need to have a sensible attitude towards the utilization of finite resources within our ecosystem is vital if the potential problems facing society are to be addressed. On the other side of our equation the nature of finite resources and our ability to meet consumption needs, without major and lasting degradation of our environment, places a major constraint on available consumption, or requires major technological leaps and/or new discoveries, or involves a radical switch in preferences so that society values the environment as a consumption good not a resource set. We can illustrate some of these points by looking at some of the economic models, and concepts, used to illuminate this area of debate. We can collapse the debate, using some acceptable limiting assumptions, to illustrate some of the key policy options; and in so doing we can highlight some of the key issues which remain unresolved.

A 'Collapsed' Neo-Classical Model

In this section we set out a 'collapsed' model which brings together some of the key elements of the neo-classical approach to environmental issues. The model is presented in order to illustrate, very simply, the essence of the economist's approach in a basic form which enables us to see the relationship between some key concepts, and to identify the assumptions on which much of the traditional economic models are built. In so doing we shall be able to see the way in which some of the conceptual issues (such as sustainable development, optimal resource use, etc.) fit together and also expose some of the more contentious elements for debate. We can then take up some of these issues later to examine the unanswered questions from the model and the appropriateness of the various assumptions, and to determine whether the economist's traditional approach is flawed, inadequate or simply incomplete.

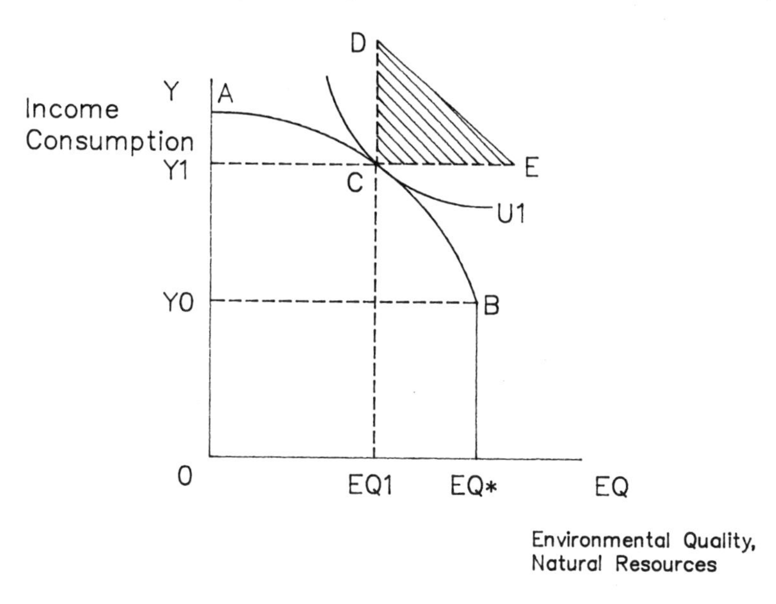

Figure 4.1

The basic components of a model for debate can be illustrated as in Figure 4.1, which brings together the fundamental elements of a system where consumers exercise choice over consumption, where the environment is valued, and where there is a fixed natural resource which comprises the environment, some of which must be transformed into goods to satisfy consumer demand. The diagram thus combines fixed environmental quality, natural resources, consumer preferences, and optimal consumption in a time period in a manner which enables us to address the issue of sustainable growth. In order to generate the equilibrium in this model we must assume additionally that there is perfect information, that property rights are clear and enforceable, that there are no transaction costs, and a given level of technology which is used to transform resources into consumption goods. Within a given fixed time period we can therefore define the optimum position for society as being at point C with consumption of Y1 and a level of environmental quality EQ1 having transformed EQ*-EQ1 resources into goods. At this point consumers in society are enjoying maximum welfare. The preferences in society, as exemplified by the indifference curve U1, are tangential to the transformation curve AB, which plots the way in which natural resources are optimally translated into consumption goods for the given technology. Sustainable development would be possible with any expansion path through time which was wholly within the shaded area DCE. This framework can be used to highlight some of the major factors affecting

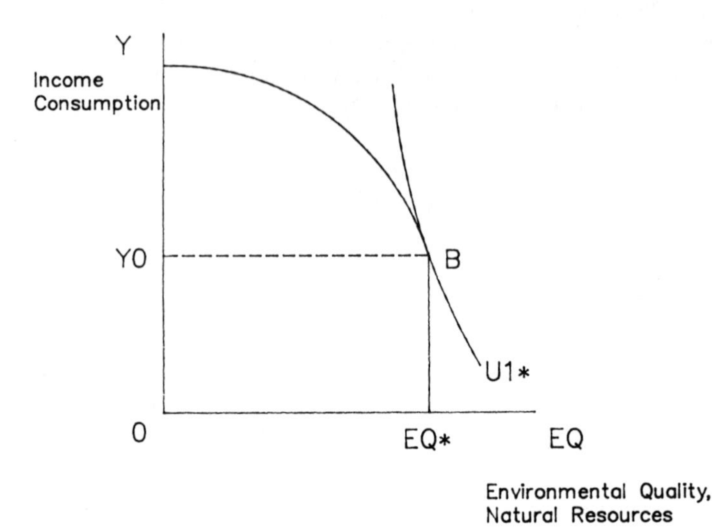

Figure 4.2

the debate on concern for long-run environmental quality and consumer choice.

If the pattern of consumer preferences was such as to lead to a corner solution at B (as illustrated in Figure 4.2) then society would be able to satisfy all its consumption needs and maximize welfare without any harmful effects on the environment. The difference between the pattern of preferences as shown by U1 in Figure 4.1 and U1* in Figure 4.2 is that the consumers as represented by U1* place a higher personal and intrinsic value on environmental quality. In valuing the environment more highly they freely choose to consume fewer products which do harm to it. They have consciously traded off higher personal consumption for higher levels of environmental quality. This simple exposition can be used to show that one policy option, education of the consuming public, if it can lead to a genuine shift in the pattern of preferences can have a positive impact on the environment. Figure 4.3 illustrates some of the variables relating to environmental objectives which may have policy implications, or be important in setting policy targets. In the figure the elements are identified by the letters a to d with the pertinent direction of desirable change shown. The precise implications are given in the Key to the figure.

It is evident that the factors outlined, namely technological advance, new discovery, and consumer attitude shifts are often cited as potential panaceas within the environment debate. However, it is also evident that for these to offer long-run solutions, major structural shifts have to take place outside the model. To examine policy options within the framework of the traditional economist's approach is to use the price mechanism (in order to equate social and private costs), to internalize externalities and effect shifts which are pareto

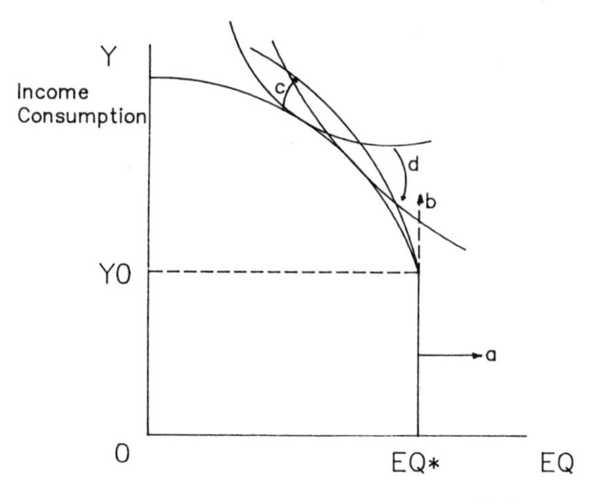

Figure 4.3

a Scientific development and new discoveries increase available resources, increase environmental quality and make the choice set wider.
b Technological advance enables more resources to be developed for consumption without harming the environment.
c Technological advance moves the transformation curve outwards by enabling more consumption goods to be produced from given levels of resources. Transformation efficiencies.
d Consumer preferences change to value environment more highly so that consumers forgo consumption goods in favour of environmental protection (or choose between goods in a complex society to select those which embody environmentally friendly characteristics).

efficient. The use of the price mechanism as a means of adjustment is in stark contrast to the judicial approach which seeks to set limits through statute law which are then enforceable through the courts. It is a matter of private regulation through prices and a fiscal regime as against external regulation through the legal system.

The price mechanism seeks to adjust equilibrium outcomes by ensuring that prices reflect external costs associated with consumption choices (the 'polluter pays' principle) (see Burrows, 1979). The implicit price of the environment, in terms of consumption, can be determined within the model from the slope of the indifference curve at its point of tangency with the transformation curve (see Figure 4.4). If we take the view that in unfettered equilibrium the consumer is engaging in levels of consumption which are detrimental to the environment, then policy-makers can influence that equilibrium through, for example, a tax regime which increases the cost of environmentally harmful consumption and thus naturally encourages the consumer to seek a new equilibrium. In Figure 4.4 the implicit price ratio is given by PEQ/PY in equilibrium at A. If a tax is introduced to discourage the consumption of goods which are environmentally

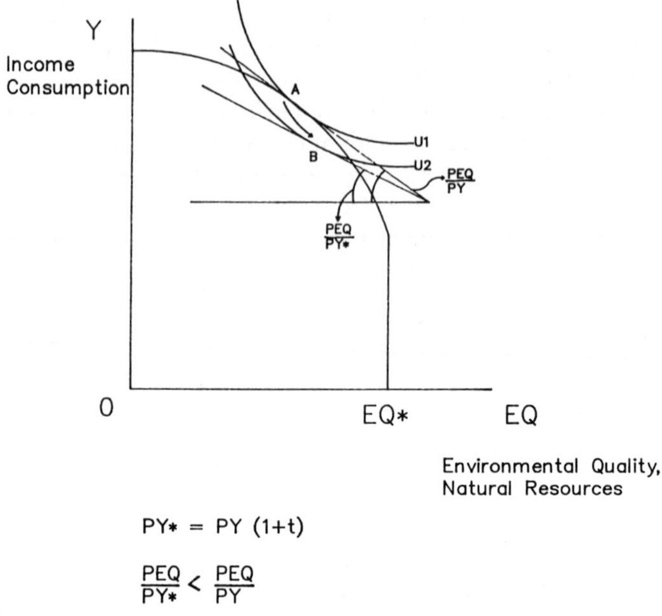

Figure 4.4

harmful (e.g. a carbon tax) then a tax will be levied on every unit of consumption (Y) so PY* = PY (1 + t) where t is the value of the tax (Pearce, 1991). As the 'price' of the environment rises so PEQ/PY* > PEQ/PY and a new equilibrium level of consumption will emerge at a level of consumption lower than that enjoyed previously and with less environmental degradation (point B).

This simple analysis can demonstrate the way a tax can be used to adjust equilibria and affect macro objectives but of course ignores, for more complex societies, the distributional effect of taxes both between time periods and between consumers within a time period. In reality, in the case of a carbon tax, the distributional consequences may be expected to be quite severe with a considerable redistribution of income from the less well off towards the better off (Symons, Proops and Gay, 1991).

Within this simplified model it is possible to illustrate a number of the issues in the environmental debate and observe why many of the popular solutions are appealing. However, it is also necessary to look a little beyond the model and examine the core assumptions on which it is based in order to simplify the treatment. We must ask whether these assumptions, singly or together, are such as to abstract so much from the real world of environmental concern that they render the model suspect. Such assumptions which are important to acts of pollution include property rights and their enforcement, transaction costs, a single time period (note the importance of inter-generational transfers, discounting and compensation), and perfect information (knowledge and rationality) (Sheldon and Smith, 1992). A further issue which may be important, but which is not taken further here, is the issue of scarcity and, specifically, the concept of economic scarcity versus absolute (see Norgaard, 1990). These assumptions underline most economic modelling at its simplest level and so, not unnaturally, may be transferred across into the arena of environmental economics. However, we need to ask whether the full breadth of environmental issues lend themselves to one simple approach or whether we need more flexibility than is contained within the conventional and unamended paradigm.

The next section looks at some of these issues in turn and illustrates the complexity behind them. Overall it will be seen that what is required is a breadth and flexibility of approach and attitudes which are prepared to accept that environmental issues often have a complexity which demands an approach from more than a single paradigm.

The Assumptions Explored

Property rights

One of the major assumptions on which much of micro-economics is based is that there are clearly defined and enforceable property rights. That is to say that private individuals or corporations have legally enforceable rights which enable them to control, transform and deploy without hindrance, those items, products or processes which they own. These rights are established and recognized in law. However, if the use of a product or the outcome of any process is such as to affect another individual or his property over which that individual also has property rights then the second individual has a right of recourse to the first. Thus if I own a pen it is mine to do with as I wish. I can sell it, transform it, destroy it or whatever. However, if I use that pen so as to infringe the rights of a neighbour the person harmed will be legally entitled to demand compensation from me and that right will be supported in law. Thus if my pollution of the environment in this context is to put an inky stain down my neighbour's shirt by negligent use of my pen he might well wish me to pay for it to be cleaned or replaced. This is a simple matter to deal with. Indeed, if I own a field in the same legal jurisdiction as a neighbour who places hazardous waste on it the matter is relatively simple to understand in the above terms: the property right is clear, and enforceable. A common construct, though not without criticisms, is the 'Coase Theorem' which establishes not only the importance of property rights but also that so long as they are clearly defined for one party the symmetry of treatment will ensure that the market reaches a social optimum (Coase, 1960). The Coase work has spawned an enormous literature of its own in the field of externalities. The environmental issues on a global scale, however, fall into an arena which is not so easy to define. Property rights may not exist, be difficult to define, or be difficult to enforce (Smith and Blowers, 1991; Clark and Smith, 1992).

In the hazardous waste example used above it was clearly easy to identify the division of property rights, in that they are within the same legal framework and therefore under the same conditions of enforceability, and in that cause and effect were easy to identify and immediate. In this local case either the use of the price mechanism (you can dump but it will cost you £x per ton) or a regulatory framework (thou shalt not dump) would be applicable and relatively non-contentious. In the context of major environmental threats, rather than small-scale local pollution, there may be no property rights (who owns the ozone layer), property rights may be ill-defined, fall into different jurisdictions (pollution does not respect national boundaries), be extremely costly to enforce, or be such as to have some doubt over cause and effect. For example, if there is a nuclear disaster in a communist state and some months later sheep in a distant country, grazing on land adjacent to both granite and a local nuclear power station are found to have higher than normal levels of radioactivity, how do you proceed? With caution, difficulty and great local, national and international sensitivity. Without labouring the point it is evident that the simple assignment of property rights which underpins many of the micro-economic approaches to pollution control, even allowing for the adaptations which have arisen through the distinction between private and social cost and the inclusion of externalities, has shortcomings in dealing with major, global environmental issues. The importance of property rights is not denied, it is the practical and universal application of the concept which is at issue. It is noticeable that in certain major problem areas one possible solution is to establish international bodies through a political process. These then assume quasi-property rights which they exercise on behalf of the parties to an agreement with an international and specific regulatory framework which signatories agree to abide by (for example, the international agreement on the reduction in CFCs).

Transaction costs

Much of the analysis in simple neo-classical models assumes that transaction costs are either zero or at least insignificant. Thus the overall equilibrium is not distorted by a disproportionate amount of resource being consumed in the transaction itself. In reality transaction costs not only exist but can sometimes be substantial. Also, the more complex the bargain, the more ill-defined the terms of the bargain, and the greater the number of parties to the bargain, the higher are the likely transaction costs. In the simple issue of the pen and the inky shirt, the cost of making a bargain to compensate is low, and the costs of establishing the property right are sunk costs in the legal framework. However, most major environmental issues are of such complexity that the transaction costs are potentially enormous. One needs to have a clear idea of cause, effect and magnitude of environmental harm before embarking on a process of amelioration because the costs of the process and the transactions are known to be so high. A major international agreement, for example, with a compensation framework relating to, say, whaling or acid rain requires enormous investment to get to a stage where any single transaction can take place. Often the transaction costs will fall unequally between parties and may tend to fall on the party that does not have a property right. In this context the Coase Theorem becomes almost redundant. If a bargain takes place the outcome is optimal, if a bargain does not take place it is because the transaction costs outweigh the potential benefit (assuming these could be accurately quantified at the outset) and therefore the outcome is optimal. The theorem becomes unfalsifiable. The existence of high transaction costs which prohibit a market solution or free bargain between the parties may provide a reason for certain types of environmental problem to be more susceptible to government regulation.

The existence of transaction costs and their distribution between parties in any potential agreement on externalities means that the outcomes are less easy to predict. Transaction costs blur the distinction between a purely economic solution, within a prevailing paradigm, and a mixed solution requiring a sensitive appreciation of a number of potential approaches.

Perfect information

The basic models also assume that information is perfect. That is to say that all parties share common and correct information about all products, their prices, their characteristics and their consequences. Thus all parties make rational economic decisions on the pattern of expenditure to generate sets of goods which perfectly meet their needs within their budget constraints. The characteristics of all these goods are known in environmental terms so that it is assumed that if two goods differ only by their environmental consequences then rational consumers could make the appropriate choice in line with their own utility function, i.e. pattern of preferences. In this way optimal decisions are made concerning patterns of preferences and as producers are able to perfectly judge those preferences they will only produce goods which fit consumer needs.

The problem of information and its imperfection in reality is the subject of enormous work and debate in economics in itself. Information can be seen as a good to be bought and sold, and the asymmetry of information in a relationship can be seen to have very distinct and often harmful consequences. In the specific context of environmental concern, it is necessary to consider this imperfection because of the generally uncertain nature of the consequences of certain actions (Houghton, 1990; Smith, 1990; Sheldon and Smith, 1992). Whilst it is true that environmentally conscious consumers cannot make rational decisions unless they are appraised of the environmental characteristics of the products open to them, it is perhaps more important that many decisions have to be taken in the absence of any information on consequences of that consumption or where some consequences may be perceived but their precise nature and/or magnitude are not known. In this world of uncertainty and imprecise information we run the risk of doing environmental damage because we simply cannot quantify the unquantifiable (Sheldon and Smith, 1992).

Thus when information is unavailable or incomplete within a given set of goods in a single known time period, or if there is incomplete information on the consequences of consumption between time periods, we cannot be certain that consumption decisions are optimal in environmental terms. There is a tendency to consume first and worry about the consequences later, rather than an unwillingness to consume. Thus we embarked upon consumption patterns locked into the burning of fossil fuels not knowing what the consequences would truly be, even though there was sufficient visual evidence from the outset (smoke, smog, etc.) to lead to a suspicion that there might be downstream consequences. However, taken to its limit the lack of information could lead to a ban on current consumption of a wide range of products. Perhaps neither extreme is sensible, but the acknowledgement that information is imperfect and that the level of knowledge varies over time should lead to a flexible, adaptive and cautious attitude to consumption, pricing and regulation over time.

Inter-generational transfers

The concept of time and the way in which decisions within one time period affect the stock of goods available for consumption in a later period and consequent levels of welfare are very important. However, the way in which some aspect of time preference and appropriate discounting can be brought into consideration is problematic. The basic premise for sustainable development is that, as a minimum condition, the consumption patterns of the current generation should leave future generations no worse off.

The traditional way by which economists deal with events which transcend time periods is to convert values to single period values through a process of discounting. Thus if an investment is made in one period which has a return in three periods time the real value of that return is calculated by discounting the future monetary value of the return by an appropriate discount rate (the prevailing rate of interest?) over the three periods to give it a value in current terms. The dimensions in this calculation are the estimated value of the return in the future period, the probability of that return, and the choice of a rate at which to discount the future return. The same procedure can, of course, be applied to the concept of future losses. In investment decisions this is the kind of calculation which is being conducted all the time. It is very susceptible to application to environmental issues. If action A will lead to loss B in x years time then what is the current value of that loss to add to the current cost associated with action A so that a rational economic decision can be made. It seems very straightforward until one examines a little more closely the nature of many of the issues of concern to us in environmental terms. The probability of certain states of the world occurring is difficult to calculate, the costs associated with certain states of the world are difficult to estimate, and the potentially devastating catastrophe where all is lost is difficult to encompass with the framework. Add to this the minor matter of a choice of discount rate and the methodology becomes much less clear cut (Markandya and Pearce, 1988). This is not a case for throwing the concept of discounting as an aid to decision-making out of the window. However, it does point again to the conclusion that it is a sensitive appreciation of use and abuse that should guide policy-makers and decision-takers not a simplistic recourse to a standard technique from a simplified model of the world.

The complexity of environmental problems

In looking a little more closely at some of the problems associated with property rights, transaction costs, inter-generational transfers and perfect information we have not invalidated the approach which economists have brought to the consideration of environmental issues; rather it has been demonstrated that the real world of environmental issues has perhaps a degree of complexity about it which embraces the need for a wider sub-set of considerations than is provided by an unthinking application of simple economic models and concepts. Whilst the simplifications can be most useful in setting certain parameters and giving guidance as to the likely direction to take, they cannot produce a simple decision system in themselves. Consequently we need to examine means by which we can incorporate such concepts without minimizing the significance of the complexity of environmental problems.

A Mixed Approach

The use of the neo-classical paradigm (or rather parts of it) to illustrate some of the issues surrounding the use of a discipline to address environmental concerns has shown that some aspects of the key environmental questions do not lend themselves easily to a paradigm that was developed initially to deal with clearly defined private market relationships. However, a sensitive appreciation of the key concepts may be an additional tool to the economist's armoury. It is important now to look briefly at what other tools you would wish to have alongside them and at how the needs of business curricula and commercial practice may be enhanced to produce a greater sensitivity and awareness in both students and managers.

A mixed approach was established as an appropriate modus operandi in the seminal text on pollution control by Baumol and Oates (1979). They introduced the concept of a 'pluralistic' approach to environmental policy where the policy-maker should have at his or her disposal means by which commercial outcomes could be influenced by a mixture of regulatory, fiscal and price controls. The target could cover prevention as well as control and compensation on output. The essential element for Baumol and Oates was that any policy should satisfy eight criteria:

1. Dependability: does it work?
2. Permanence: will it always work?
3. Adaptability: when growth occurs will the policy adapt?
4. Equity: is it fair in its burden between individuals?

5. Incentives: does it encourage maximum adjustment or minimal compliance?
6. Economy: is it cheap?
7. Political attractiveness: will politicians back it?
8. Minimal intervention: will it interfere with individual decisions as little as possible?

They could then subject certain regimes and certain policy approaches to these tests to evaluate an appropriate framework to preserve the environment. They conclude that there is undoubtedly a need for some use of price control within the mechanism, and that the economist's approach cannot be dismissed, but that a narrow paradigm view is inappropriate. What is required, they state, is 'a wide array of tools and a willingness to use each of them as it is needed.'

Thus an appreciation of regulation and alternative legal and regulatory frameworks, as well as the political and administrative context in which policy may be pursued, will be vital for the development of sound business programmes to raise the awareness of business students and prepare them to approach contemporary environmental issues with an open mind on policy. The breadth of this policy approach is covered very well in the text by David Pearce and Kerry Turner in which they note:

> Policy-makers are simply trying to discover and arrive at a set of pollution control arrangements which make enough people better off, so that those circumstances are preferable to current circumstances. ... socially acceptable policy goals should be based on the criteria of political feasibility, cost effectiveness, flexibility and equity.
>
> *(Pearce and Turner, 1990)*

Conclusions

In constructing business curricula which reflect an environmental awareness we must take note of the lessons which have been provided from this illustrative examination of one sub-discipline. First we must note that most of our basic business disciplines have been developed within a context which was not, at the outset, conscious of, or designed to reflect, environmental issues. In this respect environmental consciousness is new and the methodological challenges it poses may be complex. Second, all paradigms make simplifying assumptions in order to make their basic building block models work; the consequences of these simplifications, and their applicability when 'real world' factors are taken into account, may vary when new problems are posed for the paradigm. Third, many environmental issues are complex both in their manifestation within a time period and in their consequences between time periods. Finally, to simply adapt the tools arising from one paradigm is to deny the applicability of alternative approaches to specific aspects of common problems.

If we are now to evaluate whether the approach to curriculum should be to simply add on items within subjects to the list of factors to be taken into account or to develop new approaches which bridge boundaries then the answer would seem to be that the former would be incomplete and potentially misleading, whilst the latter would be more difficult, more demanding, but more likely to produce a comprehensive appreciation in the student. The balance of this chapter, arising out of an examination of the approach economists have taken to environmental issues, and without denying the massive contributions and literature which have been developed in this field, is to conclude that we should endeavour to develop multi-faceted environmental courses for business to address the widest range of issues with the widest set of available tools.

References

Allen, R. (1980) *How to Save the World*, Kogan Page, London.

Baumol, W. J. and Oates, W. E. (1979) *Economics, Environmental Policy and the Quality of Life*, Prentice Hall; developed in W. J. Baumol and W. E. Oates (1988) *The Theory of Environmental Policy*, (2nd edn.) Cambridge University Press.

The Brundtland Report (1987) *Our Common Future*, World Commission on Environment and Development.

Burrows, P. (1979) *The Economic Theory of Pollution Control*, Martin Robertson.

Clark, M. and Smith, D. (1992) Paradise lost? Issues in the disposal of waste, in M. Clark, D. Smith and A. Blowers (eds.) *Waste Location: Spatial Aspects of Waste Management, Hazards and Disposal*, Routledge, London.

Coase, R. H. (1960) The problem of social cost, *Journal of Law and Economics*, October.

Coyne, J., Smith, D. and Hart, D. (1991) Greening the business school: source of change or bastion of resistance? Paper presented at the First Annual European Environment Conference, University of Nottingham, September.

Croomer, J. (1979) The nature and quest for a sustainable society, in J. Croomer (ed.) *Quest for a Sustainable Society*, Pergamon Press, Oxford.

Houghton, J. (1990) Scientific assessment of climate change. The Policymaker's Summary of the Report of Working Group 1 to the Intergovernmental Panel on Climate Change, WMO/UNEP, July; and the reply and critique to it: George C. Marshall Institute (1990) Scientific perspectives on the greenhouse problem, George C. Marshall Institute, Washington DC.

The Lancet editorial, Vol. 336, 15 September 1990.

Markandya, A. and Pearce, D. W. (1988) Environmental considerations and the choice of discount rate in developing countries, *Environment Department Working Paper No. 3*, World Bank, Washington DC.

Norgaard, R. B. (1990) Economic indicators of resource scarcity: a critical essay, *Journal of Environmental Economics and Management*, Academic Press, July.

Pearce, D. W. (1991) The role of carbon taxes in adjusting to global warming, *Economic Journal*, Policy Forum.

Pearce, D. W. and Turner, R. K. (1990) *Economics of Natural Resources and the Environment*, Harvester Wheatsheaf.

Sheldon, T. A. and Smith, D. (1992) Assessing the health effects of waste disposal sites: issues in risk and analysis and some bayesian conclusions, in M. Clark, D. Smith and A. Blowers (eds.) *Waste Location: Spatial Aspects of Waste Management, Hazards and Disposal*, Routledge, London.

Smith, D. (1990) Corporate power and the politics of uncertainty: risk management and the Canvey Island complex, *Industrial Crisis Quarterly*, Vol. 4, No. 1.

Smith, D. and Blowers, A. (1991) Passing the buck: hazardous waste disposal as an international problem, *Talking Politics*, Vol. 4, No. 1.

Smith, D. and Hart, D. (1991) Business education and the environment: ramshackle or innovative? Working paper, Liverpool Business School.

Symons, E. J., Proops, J. L. R. and Gay, P. W. (1991) Carbon taxes, consumer demand and carbon dioxide emission: a simulation analysis for the UK. Paper presented at the First Annual European Environment Conference, University of Nottingham, September.

5

THE EMERGING GREEN AGENDA: A Role for Accounting?

Dave Owen

During the 1980s, with ever increasing emphasis being placed on financial efficiency and value for money as representing the keys to social prosperity, accounting inevitably came to occupy a position of prime importance in the public policy arena. As Hopwood (1984, p. 168) puts it,

> Accounting has ... been implicated in a more positive shaping and influencing of that which is regarded as problematic, the forms which public debates take and the options seemingly available for management and public action.

An overriding concern with accounting notions of profit and loss has indeed permeated debate over key issues such as privatization and de-industrialization, the latter perhaps most graphically illustrated during the course of the 1984–5 mining dispute (see Cooper and Hopper, 1988).

However, by the end of the decade, alternative 'bottom line' measures to that of purely private profit had begun to achieve a measure of prominence as major reservations were increasingly expressed about the impact of economic activity on the quality of life, evidenced by concern over issues such as natural resource depletion, unacceptably high levels of pollution, global warming, acid rain and deforestation. The success of environmental pressure groups in ensuring that these wider ramifications intruded into public consciousness, and ultimately the political agenda, together with the increasing influence of green consumerism and the emergence of a green investment movement, has elicited a significant response from the business community. Not only have an ever growing number of companies felt the need to parade their green credentials, but also prominent business organizations such as the Confederation of British Industry, the British Institute of Management and the Institute of Directors have launched a series of green initiatives. Furthermore, environmental consultancy is bidding fair to replace value for money and efficiency studies as the growth area of the 1990s.

Such developments would seem potentially to have considerable repercussions for the accounting function, particularly should companies come to view their annual report and accounts as a major medium of communication in respect of environmental and social issues, and management information systems are developed which are designed to promote the green dimension to a central position within the corporate decision-making process. Indeed, the President of the Institute of Chartered Accountants in England and Wales has recently gone on record as stating that,

> In responding to the challenges posed by the environment, which is our natural wealth, all aspects of accountancy including financial reporting, auditing, management accounting and taxation will have to change. In doing so, there will be an impact on all members of the Institute whether in public practice or in commerce and industry and whether working at home or abroad.
>
> *(Lickiss, 1991)*

The plethora of recent articles on green accounting issues in professional journals, and the active involvement of the major professional accounting bodies in terms of funding research and establishing working parties, together with the strenuous efforts of leading international accounting firms to corner the environmental audit market all tend to add credence to this particular view.[1]

A Role for Accountants?

Whether or not one applauds the attempts of the accounting profession to jump on the green bandwagon largely depends upon which of two widely diverging perspectives, which I would term 'radical green' and 'reformist', one adopts concerning the implications for the accounting function of the continually evolving green agenda.

The 'radicals'[2]

From the radical (or 'dark green') perspective accounting may be considered as being a craft, or discipline, which is simply reflective of our current economic and social system – a system obsessed with the necessity for economic growth. In a particularly penetrating analysis the French philosopher Andre Gorz, for example, points out that

> once you begin to measure wealth in cash, enough doesn't exist. Whatever the sum, it could always be larger. Accountancy is familiar with the categories of 'more' and of 'less' but doesn't know that of 'enough'.
>
> *(Gorz, 1989, p. 112)*

In other words, accounting is unable to cope with notions such as 'sustainable' or 'sufficient' and therefore is of very little relevance to the central issues raised in the current green debate. The fundamental concern in Gorz's analysis, then, becomes one of clearly defining limits within which economic rationality (and associated accounting techniques) is to operate, rather than permitting its seemingly never-ending expansion, and to reclaim other areas of human activity as the preserve of moral or aesthetic criteria. The call for a spiritual re-awakening is indeed central to much radical green analysis (see, for example, Porritt, 1984 and Dobson, 1990), a point clearly recognized in critiques of traditional economic thought such as those provided by Schumacher (1973) and more recently Daley and Cobb (1990) which locate current environmental problems in the spiritual failings of western society and offer a new paradigm for economic and public policy issues underpinned by an insistence on the very need for a spiritual re-awakening.

An additional point of departure for a prescriptive analysis of society's current environmental and social malaise from a radical ecology perspective lies in its rejection of the pursuit of economic growth as an essential societal goal. This position is neatly encapsulated in Porritt's comment that

> if you want a simple contrast between green and conventional politics, it is our belief that quantitative demand must be reduced, not expanded.
>
> *(Porritt, 1984, p. 136)*

Achievement of such an end entails making a sustained assault on the market system (Ryle, 1988; Gorz, 1989; Seabrook, 1990) which means going beyond a mere critique of the rules of economics, or indeed accountancy, and requires nothing less than the invention of new institutions for a different, non-market, economy (Ryle, 1988). Unfortunately, the radical ecologists' critique doesn't extend to the elaboration of new 'accounting' systems which would operate in a non-market economy. However, there is a clear denial of the efficacy of applying principles of neo-classical economics, via market-based incentives, towards solving environmental problems (Pearce *et al.*, 1989) or indeed of seeking to develop an accounting response within this particular framework, as one recent highly influential study has sought to do (Gray, 1990a).

In sum, the above analysis, although of necessity being somewhat brief, clearly indicates that from the perspective of radical ecology the environmental and social arena is not the place for economic rational (wo)man and certainly not accountants!

The 'reformists'

The alternative, reformist perspective, by contrast, noting accounting's newly won position of prime importance in the public policy arena (an observation made at the commencement of this chapter) suggests that, like it or not, accounting information is therefore vitally important in moulding perceptions of what constitutes good or bad organizational performance. Accounting's strength lies in its ability to make visible and discipline performance (Hopwood, 1984) but in focusing on issues of profit and efficiency, whilst ignoring social and environmental concerns, conventional accounting techniques are heavily implicated in the current environmental mess we've got ourselves into. Interestingly, a similar observation was made by John Maynard Keynes some sixty years ago when he noted that, 'Under the peculiar logic of accountancy, the men of the nineteenth century built slums rather than model cities because slums paid' (Keynes, 1933). The reformist, therefore, goes on to argue that there is a pressing need to reform accounting practice so that the wider aspects of performance are captured, and hence enter into the decision-making process.

Such a motivation underpinned the efforts of accounting researchers in the 1970s who sought to develop workable social accounting performance measures and reporting techniques (see Gray, Owen and Maunders, 1987). Essentially the approach may be labelled technocentrist, in contrast to the ecocentrist position outlined earlier, in that it pursues a belief in the ability and efficiency of management to solve environmental and social problems by objective analysis based on the provision of 'better' information. Furthermore, a desire is expressed to accommodate such problems within the prevailing socio-political system, utilizing a free market economy approach or via gradualist liberal reform of the market system (Pepper, 1984). Little challenge is therefore posed to prevailing economic orthodoxy, perhaps most notably propounded by the Brundtland Commission (World Commission on Environment and Development, 1987), suggesting that industry is both central to the economies of modern societies and an indispensable motor of growth.

The reformist approach is clearly therefore open to the accusation, from those of a more radical persuasion, of exhibiting a passive acceptance of the existing social and political context of corporate reporting (Cooper and Sherer, 1984), so that any prescriptions derived for changes in accounting practice may be considered as merely an exercise in immanent legitimation (Puxty, 1986). As Tinker *et al.* (1991, p. 29) put it, such 'middle of the road' theorizing is

> prompted by concerns about what is politically pragmatic and acceptable; not what is socially just, scientifically rational, or likely to rectify social ills arising from waste, exploitation, extravagance, disadvantage or coercion.

The sum result therefore is that the palliatives thereby advanced deal with symptoms – not causes.

Exploring the Reformist Perspective

Whilst acknowledging the strength of the above radical critique of the reformist approach my intention in the remainder of this chapter is nevertheless to consider the potential implications for the accounting function of adopting such a perspective. My reasons are twofold:

1. Whilst radical ecology provides us with a fundamental critique of current patterns of consumption and production, very little serious thinking seems to have gone into developing actual strategies for change (Frankel, 1987; Dobson, 1990).[3] There are strong grounds for believing that our current environmental

and social malaise is of such a magnitude that abstract theorizing provides an insufficient response, and that by at least starting to move away from the exclusive emphasis on short-term financial performance prevailing at the moment in western economies the reformists do begin to point a way towards *practical* change.

2. The reformist approach is undoubtedly more in tune with influential current trends, perhaps most notably encapsulated in the Pearce Report's emphasis on financial quantification and market-based incentives. That accounting has a major role to play in advancing such trends is evidenced by the recently enacted Environmental Protection Act's espousal of the concept 'best available techniques not entailing excessive cost' (BATNEEC) in seeking to modernize pollution control together with the emphasis on cost effectiveness in sundry Department of Trade and Industry publications concerned with 'business and the environment'. More fundamentally, as Gray (1990a) points out, further implementation of ideas put forward in the Pearce Report will probably not be possible without both a voluntary response from organizations and, especially, some new environmental accounting systems to support them.

   From a reformist perspective the ever developing green agenda has profound implications for accountants and the accounting function, both in terms of developing internal management information systems and external reporting practice. It is to a consideration of these two distinct issues that we shall now turn our attention.

Management information systems (internal reporting)

From the foregoing remarks it seems clear that the accountant has a role to play in the initial stages of the 'greening' of organizations in developing information systems capable of capturing the cost effects associated with the adoption of environmentally friendly practices. However, it would be unduly restrictive to see the accountant's role solely in these terms.

Indeed, Gray (1990a) has argued that accountants tend to overestimate their abilities in the realm of attaching financial numbers to various aspects of business activity whilst, at the same time, probably underestimating their most important talent – that of the design of, recognition of, assessment of and control of the information systems in an organization. Much recent discussion has centred on how best accountants may employ these latter talents in the development of what is becoming a prime tool for managing corporate environment performance, the environmental audit. Indeed, one commentator has recently suggested that

> In order for credible environmental audits to take place the process must borrow wholesale from the experience and expertise which accountants and the accounting/auditing profession have developed over the years.
>
> *(Adams, 1991, p. 77)*

Environmental audits

Environmental auditing has been defined by the International Chamber of Commerce (1989) as being

> A management tool comprising a systematic, documented, periodic and objective evaluation of how well environmental organization, management and equipment are performing with the aim of helping to safeguard the environment by:
> (i) facilitating management control of environmental practices;
> (ii) assessing compliance with company policies, which would include meeting regulatory requirements.

Recent surveys published by the World Wide Fund for Nature (Elkington, 1990) and *Green Magazine* (Grant, 1990) indicate the considerable extent to which leading British companies are beginning to employ environmental audit techniques, with names such as Proctor and Gamble, ICI, Shell and Lever Brothers figuring prominently.

Ideally, an environmental audit will entail the carrying out of a 'cradle to grave' assessment of how an organization's activities affect the environment. In the case of a manufacturing company, for example, the assessment would cover:

- extraction of raw materials used in the production process, particularly taking account of the depletion of non-renewable resources.
- direct impact of the production process, in terms of emissions, waste and land use, with particular attention being paid to the adequacy of risk management techniques for dealing with the dangers of environmental accidents.
- use and final disposal at the end of their useful life of the organization's products.

Finally, the audit report ultimately produced should constitute not only a formal statement concerning the current status of compliance with statutory and other, possibly internally derived, requirements but also, more proactively, a programme for future action (Maxwell, 1990).

Blaza's (1991) description of the integrated environmental review recently carried out by the UK subsidiary of Norsk Hydro, a major industrial conglomerate with a significant presence in the fertilizer, PVC polymer and aluminium industries amongst others, covering the operations of the entire organization provides an excellent example of the practical application of such cradle-to-grave audit procedures.

> Each of the manufacturing companies was reviewed, firstly in terms of performance against the appropriate regulations. Both employee health and safety matters and the effects on the external environment were covered. The group's main products were subjected to an 'eco-balance' review – the raw material and energy inputs required to manufacture them, their environmental impact in use, comparisons with alternative products and finally any potential problems after use, including waste disposal and potential for recycling.
>
> The complete report once finished was subjected to review by an authoritative independent consultancy whose remit was to confirm compliance with the environmental, health and safety regulations, to validate individual reports and data, to

investigate environmental practices on site and to make recommendations for change as appropriate.

(*Blaza, 1991, p. 207*)

In practice there is a wide range of different types of environmental audit that can be carried out. Elkington's (1990) discussion of current practice at British Petroleum illustrates this point, although it should be noted that in each case the approach adopted follows a common pattern based on gathering information by interview, site inspection and examination of relevant documentation.

The different types of audit currently used at BP include:

- Compliance Audits – relatively straightforward, albeit time consuming, and covering compliance not only with statutory obligations but also industry-level voluntary codes and internally generated corporate standards.
- Site Audits – the carrying out of spot checks at sites having actual or potential problems.
- Activity Audits – evaluating policy implementation in respect of activities that cross business boundaries (for example, group shipping operations).
- Corporate Audits – conducting an audit of an entire BP business sector. The aim is to ensure that roles and responsibilities are fully understood, technical and advisory support is available and all relevant communications channels are operational.
- Associate Audits – auditing companies acting as agents in overseas markets.
- Issues Audits – focusing on specific key environmental issues (for example, impact on tropical rain forests) and involving an evaluation of policy, guidelines, operating procedures and actual practice within the organization as a whole.

Essentially, environmental auditing methodology is largely derived from longer established operational (or internal) and external compliance audit processes developed and refined over many years by the accounting profession. Indeed, as Jenkins (1990) points out, in outlining the approach to environmental auditing adopted by the major international accounting firm Coopers and Lybrand Deloitte, just as is the case with the more traditional financial audit, the environmental auditor operates most effectively by tracking laid down management procedures and assessing both whether they are likely to achieve their objectives and whether they are being followed. Thus it is by virtue of their experience and expertise in the realm of information systems and control procedures generally that accountants can make a considerable contribution in the environmental auditing field, notwithstanding their lack of detailed technical and scientific knowledge in respect of the environmental issues themselves.

Whilst the overwhelming majority of the literature on the development of internal environmental accounting and information systems, and more particularly descriptions of practical corporate initiatives undertaken, has focused on environmental auditing and review techniques, there are a number of other ways in which the accounting function can contribute towards organizations becoming more environmentally sensitive. Gray (1990a) provides a particularly comprehensive analysis of possible developments in management information systems, of which the following in particular appear to present potential practical opportunities.[4]

Energy audits and energy accounting systems

Recognition of the need to make Britain more energy efficient has been a central feature of our heightened awareness of green issues. For many manufacturing companies energy costs are a highly significant element of their overall cost structure, a factor which has led to the undertaking of systematic energy audits which seek to account for energy expended in the provision of goods and services and, more particularly, to pinpoint inefficiencies and consequent saving opportunities. Interestingly, Gray points to the fact that because energy can be reduced to universal measures (ergs, joules, etc.) some attempts have been made in the past to mirror a traditional accounting system with an energy accounting system, and that whereas these ideas had little impact at the time they may well be due for re-examination. Such a re-examination would undoubtedly be encouraged should much mooted carbon taxes be introduced and energy prices thus inevitably increase substantially.

Environmental Impact Assessments

Increasingly, as EC regulations become ever tighter, environmental considerations are coming to feature highly when new projects necessitating planning applications being made are under review. A major aim in carrying out Environmental Impact Assessments is to evaluate the total possible impact a new project will have so as to be able in particular to predict whether the organization is likely to fall foul of environmental planning regulations, or indeed pressure groups. To the extent that this happens the greater the costs that will be generated in terms of delays incurred, management time wasted and compensation payments necessitated. The assessment enables the organization to compute these likely costs and would also include a consideration of relative costs attached to other project options. Amongst major companies currently carrying out Environmental Impact Assessments as a matter of course are Shanks & McEwan, the leading waste management company in the UK, and BP, where the assessments form an integral part of the company's Environment Protection Management (EPM) Policy (see Blaza, 1991).

Environmental budgets

In order to integrate environmental awareness fully within traditional financial and marketing objectives one needs to be able to rank environmental criteria on some comparable level with the more traditional performance measures employed. One possibility is to allocate levels of environmental activity along with other levels of budget allocation to activity centres and, furthermore, to tie in an element of the reward and penalty system to the satisfaction of allocated budget level. Interdivisional transfers could be dealt with, in principal, by a system of financial charges (perhaps tradable pollution permits) or the budgetary system kept in physical units and environmental interdependencies recorded accordingly. As Gray notes, very few experiments in this area have reached the public domain and there is an urgent need for in-depth field research to examine the feasibility and practicability of such ideas.

Environmental hurdle rates for new investments

The conventional approach to discounting, whereby uncertainty is handled by employing short-term payback criteria and inflated discount rates, arguably discriminates against giving a fair weighting to environmental factors. For example, the future cost of safe disposal at the end of an asset's useful life may not be taken fully into account, whilst environmental projects with long gestation periods and low values in current prices, for example reforestry, are undervalued (Hackett, 1991). Using lower discount rates for particular environmental benefits would ensure that a significant present value would attach to them, even if they were to occur many years in the future. Alternatively it may be necessary to build qualitative criteria into investment hurdle rates. The Environment Protection Act's espousal of the BATNEEC concept[5] certainly indicates that organizations will need increasingly in the future to provide both qualitative and quantitative analysis of a chosen project's environmental impact and be able to justify 'hurdle' rates used. Thus one can anticipate accountants becoming increasingly concerned with the theoretical and applied development of investment appraisal methodologies capable of incorporating environmental factors.

The above outline of merely a few possible developments in internal reporting mechanisms, whilst being of necessity very sketchy, perhaps indicates the not inconsiderable role accountants may play in 'greening' business organizations as well as the complexity of the issues involved. Other potential information systems developments raised in Gray's analysis, most particularly establishing mechanisms for monitoring the maintenance, enhancement and depletion of natural capital (particularly 'critical' natural capital, for example the rain forests, which we expend at our utmost peril), introduce further complexities still. At the present time there is unfortunately a dearth of information in the public domain concerning practical applications of many of the suggested techniques. Thus, as Gray argues,

> There is a most urgent need for all these techniques to be explored, for experience to be shared and for experiments and research to be undertaken if we, as accountants, are to be able to contribute anything significant to the greening of organizations.
>
> *(Gray, 1990a, p. 103)*

However, there is also arguably an equally, if not more, urgent need for business organizations to convey information concerning the environmental and social impact of their activities to a wider audience via the external corporate reporting function. It is to this issue that we shall now turn our attention.

External reporting[6]

Within the business community it appears that support for environmental initiatives is often accompanied by a profound reluctance to make detailed information on corporate environmental impact publicly available. Thus the International Chamber of Commerce (ICC), whilst expressing full support for the adoption of environmental auditing programmes by business organizations has, at the same time, seen fit to stress their role as being one of a purely internal tool whose findings are exclusively for the use of corporate management (ICC, 1989). This view has been echoed by a leading member of the UK auditing profession, Brian Jenkins, Head of Audit at Coopers and Lybrand Deloitte, who suggests that in many ways the issues covered in environmental audit reports are too 'soft' for a proper form of external public reporting and that inevitably such reports will descend to an unimaginative statement on compliance with rules rather than genuinely trying to add value (Jenkins, 1990).

Such views as those outlined above dismiss too readily the role of external reporting as a vital mechanism of corporate accountability. Admittedly, Jenkins is supportive of some degree of public disclosure, in particular the publication of a statement of environmental policy within the directors' report. However, a major constituency to which the company is legally accountable, namely investors, is increasingly calling for more detailed information than that provided by vague policy statements. This desire to be kept fully informed on the environmental, and indeed wider social, impacts of company performance is, significantly, not confined to the newly emergent 'ethical' investor (see Harte *et al.*, 1991), but is also increasingly being expressed by the more traditional, solely profit-seeking investor (Cooper and Filsner, 1991).

Important as investor needs may be, it would, however, be arguably unduly restrictive to focus sole attention on that particular constituency when considering the issue of corporate accountability for environmental and other social impacts. As Gray (1990a, p. 104) puts it,

> The environmental debate has raised a long stated concern that the race, the rest of life and future generations have significant rights to information about those things which may well affect their continued existence and will certainly affect the form and quality of that existence.

Such a view recalls arguments for the adoption of the concept of public 'accountability' put forward by the authors of *The Corporate Report* some years ago (ASSC, 1975). Essentially, their suggestion was that there is an implicit responsibility incumbent upon economic entities regarded as significant in terms of their scale of command over human and material sources such that their activities have significant economic, and one might add here environmental, implications for the community as a whole, to report publicly the effects of their actions.

The foregoing analysis highlights the potential importance of external company reports as a means of accountability in a purportedly democratic society, with the capability, in particular, of influencing perceptions of business performance in the widest sense of the term. It would therefore be of some interest to ascertain how far UK domiciled companies seem willing to accept such a wider accountability for their actions in responding to the emerging green agenda via their external reporting practices.

Current UK corporate practice

Surveys carried out in the 1980s clearly indicated that UK companies, in general, attached a low priority to the voluntary reporting of *social* information[7] (Gray, Owen and Maunders, 1987; Guthrie and Parker, 1989; Gray, 1990b). Relatively few companies devoted more than one page to such disclosure and hardly any were prepared to make 'bad' news public. Furthermore, only human resource information appears to have been provided with any degree of consistency, although there is a discernible trend in recent years towards more provision of consumer orientated and community involvement data. Finally, and most significantly in the context of this particular chapter, none of the studies was able to point to much interest on the part of UK companies in disclosing details relating to the environmental impact of their activities.

Four further, more recent, studies, conducted when the recent major explosion of interest in green, and more particularly environmental, issues was well under way, and concerned specifically with corporate attitudes to environmental issues rather than social reporting in general, somewhat disappointingly back up the latter finding.

Touche Ross (1990)

An in-depth study conducted by Touche Ross Management Consultants during the latter part of 1990 into attitudes towards environmental issues on the part of thirty-two major UK companies reported that:

- Whereas more than half the companies studied claimed to devote some coverage to environmental issues in their annual report, only a few dealt with the issues in any depth, with a mere handful devoting as much as half a page specifically to environmental matters.
- The continued human resource slant to reporting is evidenced by the fact that the frequently mentioned issues are health and safety at work and working conditions. (However, there is a discernible trend towards giving some prominence to other issues, notably control of emissions, conversion of company cars to lead-free petrol, elimination of CFCs from products and product safety.)

Institute of Business Ethics (1990)

The Green Alliance, on behalf of the IBE, conducted a questionnaire survey amongst the chief executives of 500 businesses listed in *The Times* Top 1,000 early in 1990 which was designed to assess how far individual companies are developing specific policies and programmes for improving environmental performance. Eighty-two full responses and a further five partial responses were obtained. On the specific issue of whether details of environmental performance were set out in the annual report and accounts only twelve companies replied in the affirmative, whilst a further sixteen indicated that this was an issue under consideration. Additionally a few respondents, the number of which is not quantified in the published study, intimated that the information was made available in other ways.

Coopers and Lybrand Deloitte (1990)
Gallup was commissioned to undertake a telephone survey to discover, amongst other issues, the extent to which environmental matters were considered a concern for business, and more particularly the finance function. Whereas the vast majority of the 108 respondents (randomly selected from *The Times* Top 1,000 list and interviewed between 28 August and 3 September 1990) saw the environment as a significant issue for their business, on the specific matter of disclosure a mere 29 per cent of companies had ever included environmental information in their annual report. Some commitment to a change in policy is evident in that only 39 per cent expected to ignore the issue in their next report, although, significantly, very few envisaged including a page or more of information or indeed providing a separate report.

Tonkin (1991)
In 1991, after an eight-year gap, the annual survey of UK published accounts, *Financial Reporting*, published under the auspices of the Institute of Chartered Accountants in England and Wales, re-introduced some slight coverage of social accounting issues in devoting three pages to the response towards environmental issues in annual reports. From a sample of 100 large listed companies, 12 per cent highlighted environmental protection in a separate statement whilst a further 11 per cent mentioned the issue in the context of some other, more general, statement. The most common form of disclosure was simply a general statement of good intent with only 6 per cent disclosing detailed environmental policies and just 3 per cent costs incurred. The remaining 77 per cent of annual reports studied gave no disclosure, or evidence, of environmentally sensitive activities. A startlingly high figure of 96 per cent of a further sample of 150 medium-sized listed companies and 100 per cent of 50 large unlisted companies also fell into this latter category.

The overall impression conveyed by the surveys we have just considered is one of a very limited response on the part of UK companies towards the evolving green agenda in terms of making salient information publicly available. However, there is also a suggestion that a few companies at least are beginning to tackle green reporting issues with some degree of rigour. A study by Harte and Owen (1991) attempts to provide a flavour of these more innovative approaches by considering specific examples of current reporting practice drawn from a sample of thirty companies singled out as 'good' disclosers.[8] Harte and Owen's analysis is based on the latest available annual report for each company as at the end of June 1990, together with, where possible, the previous year's report for comparative purposes. In highlighting the activities of companies at the leading edge of green reporting, some pointers are perhaps provided as to the direction of future corporate reporting practice in general.

Firstly, and most significantly, a clear trend is apparent towards giving more exposure to green, and particularly environmental, issues on the part of the companies studied over the two-year period. The overwhelming majority devote more space to coverage of green issues and in a number of cases the increase is significant.

Two examples of innovatory practice are worthy of particular mention. Firstly, Caird Group, a company operating in the waste disposal industry, broke new ground in its 1989 published accounts by presenting an independent environmental audit report prepared by a major engineering and environmental consulting group. The audit covered all activities carried out at a representative sample of the company's sites and included matters relating to health and safety on-site and potential environmental impacts off-site. Secondly, RPS Group, a relatively small environmental consultancy quoted on the Unlisted Securities Market, introduced a three-page statement of environmental policy objectives in its 1989 annual report. The statement is particularly noteworthy in not only listing in general terms the key policy areas (energy; waste; transport; communication, education and training; environmental awareness; purchasing standards and commitment to environmental auditing) and aims of the company, but also describing in some detail the methods to be employed in applying the policy to corporate activities.

Turning to the other companies appearing in the sample, two disclosure trends are particularly discernible. Firstly, a marked tendency is noted towards devoting increased coverage to environmental issues in the (unaudited) 'Review of Activities' section of the annual report. Secondly, there is a trend towards introducing specific comment on environmental issues within the Chairman's statement. This comment generally takes one of two forms, either drawing attention to the opportunities for profitable trading presented to the company by increasing levels of green awareness or, somewhat less prevalently, placing on record what might be termed a commitment to good citizenship.

Overall, a wide variety of techniques were employed by the selected companies for disclosing items of social and environmental information. Perhaps particularly noteworthy were Body Shop's publication of a separate 32-page report, *Another Year Into Our Lives*, covering a range of employee, consumer, community and environmental issues and ICI's inclusion of a ten-page feature, 'World Solutions', within their annual report comprising case studies of particular individuals whose quality of life had purportedly been improved through use of the company's products. However, two particular disclosure techniques seem to be gaining more widespread acceptance:

1. Twelve companies (40 per cent of the sample) utilized what may loosely be termed a Statement of Corporate Objectives in drawing attention to their commitment to the pursuit of social and environmental, in addition to profitability, goals. In each case this comprised a statement of general

philosophy or policy, very much along the lines of one of the recommendations of *The Corporate Report* (ASSC, 1975), although a further refinement suggested in the latter document, namely inclusion of quantified information concerning medium-term strategic targets, did not find favour. Typical examples are provided by Rentokil's statement that 'Services are committed to improving the environment and protecting health and property', and ICI's declared aim to 'operate safely and in harmony with the global environment'. A notable feature in general of annual reports containing a statement of Corporate Objectives is that they then go on to devote considerable attention elsewhere in the annual report, often in the order of at least three or four pages in total, to their pursuit of the environmental and social goals for which commitment has been pledged.

2. 'Specific narrative' is clearly emerging as an increasingly prevalent disclosure style. This may prove to be a particularly important development as specific narrative, unlike general statements of good intent, is potentially auditable information.

Tesco's 1990 annual report provides a good example of this approach. Amongst other issues addressed in a separate environmental review statement it is disclosed, for example, that,

> During the year we eliminated CFCs from all products.
>
> Facilities for recycling one or more materials are provided at all our new stores. By the end of the year, recycling facilities had been installed at 135 of the 160 stores where there is space for them.
>
> 45 per cent of our petrol sales are now unleaded. The national figure is 25 per cent.

Further refinements in terms of specific narrative disclosure relate to the provision of financially quantified information and, perhaps of more fundamental importance, the introduction of an external reference point by which performance may be judged. In the case of the former, whereas intrusion of financial data is fairly prevalent in the human resource and community involvement areas of reporting, it is far more of a rarity in discussion of environmental impact. ICI is very much an exception in disclosing the total amount spent on environmental protection, in addition to making specific provision in its accounts for future clean-up costs, whilst a handful of other companies, for example Glaxo and British Gas, provide isolated examples of specific environmental expenditures. With the possible exception of mentioning the receipt of awards from outside bodies in recognition of corporate achievements in the field of environmental protection or health and safety at work, the introduction of external reference points, which may be loosely termed a 'compliance with standards' approach to social and environmental reporting, is again very much the exception rather than the rule. Amongst companies going down this route are Simon Engineering which notes the aim of all companies in the group to obtain the British Standard BS5750 qualification (a recently introduced national standard for quality systems) and BP which discloses that 'Audits based on the International Safety Rating System (ISRS) were completed and follow-up inspections started in our major operations. Four of our refineries achieved particularly high ratings under the demanding ISRS standards.'

Pressures for change

I have, in the above analysis, attempted to convey at least a flavour of how the more innovatory companies are tackling the issue of 'green' reporting and hence, perhaps, beginning to demonstrate a wider accountability in terms of their environmental and social impact. However, it must be borne in mind, in view of the survey evidence presented earlier, that such companies are very much in the minority. Indeed, a further note of caution is called for in that, even in the case of 'better' disclosers, information provision is generally highly selective and largely public relations driven with a virtual universal reluctance to disclose 'bad' news. Of course, this discerned tendency to err on the side of self congratulation merely invites an increasingly cynical response from any intended audience. I would, however, anticipate that encouragement for a more rigorous and less partial approach will be forthcoming from a variety of sources, in particular the following.

Domestic legislative developments

Whilst no new statutory standards appear to be immediately on the horizon, the Environmental Protection Act and subsequent White Paper have demonstrated the beginnings of a commitment to greater openness. Pressures for extending this openness into the formal corporate annual reporting mechanism are already being felt. For example, the Labour Opposition tabled an amendment to the 1989 Companies Bill during its committee stage which would have required companies to explain their policy regarding control of pollution and report any fines or other penalties incurred by the company in respect of breaches of legislation. More recently the Campaign for Freedom of Information and Citizen Action Compensation Campaign have been actively lobbying for a private members bill to force companies to disclose in annual reports details of environmental and safety offences and compensation payments to accident victims.

Supranational influence

Developments within the European Community are, of course, having an ever increasing influence on UK business practices. It should be noted, therefore, that the Single European Act makes it clear that environmental considerations are to be fully integrated into moves towards establishing an internal market. Furthermore, there are already more than 200 extant Directives concerning the control of air and water pollution, and recently a draft Directive calling for the environmental auditing of companies whose activities have a significant impact on the environment has been circulated for discussion.

Perhaps more remotely in terms of immediate practical implementation, the United Nations Intergovernmental Working Group of Experts on International Standards of Accounting and Reporting (a Committee of the UN's Economic and Social Council) have turned their attention to information disclosure relating to environmental measures. Amongst their initial suggestions is a call for companies to disclose:

- environmental policy
- environmental expenditure in the accounting period
- future expenditure (contingent liabilities together with voluntary and semi-voluntary clean-up and restoration costs)
- activity and performance (for example, emission levels, noise levels, toxic waste, etc.) measured against statutory and internally generated targets.

The issue of environmental disclosure is, furthermore, now firmly established as a continuing area of interest for the group.

Influential user groups

Recent years have seen the establishment of a thriving UK social/green investment movement, of which the ethical and environmental unit trusts are a rapidly growing component part, which is increasingly calling for accounting policy-makers to give a higher priority to issues of environmental and social disclosure. There are, furthermore, some signs that the concerns of the social investor are eliciting a wider institutional response, a development that mirrors previous American experience (see Harte *et al.*, 1991).

Corporate self-interest

This perhaps represents the most immediate pressure for change. Not only is rigorous disclosure the most potent defence against corporate critics (for example, the Social Audit Movement – see Gray, Owen and Maunders, 1987 and Geddes, 1991), but also, as increasingly onerous environmental legislation imposes ever higher financial costs on companies, it seems eminently sensible to report fully in order to justify expenditure incurred.

Should the above pressures for change be as great as I suspect, one might again, just as was the case for internal reporting considered earlier, expect there to be major implications for accountants who have traditionally been at the forefront of developments in external corporate reporting practice.

In Conclusion: Some Potential Pitfalls

My main task in this chapter has been to outline, from what I have termed a reformist perspective, the potential implications for the accounting function of current trends towards heightened levels of green awareness throughout society. It has been suggested that the accountant can make a, not inconsiderable, contribution in two areas:

1. Developing management information systems to assist organizations in

responding at a micro level to the macro level initiatives introduced following the Pearce Report.
2. Developing external reporting practice in order to promote public accountability, the need for which is clearly central to the emerging green agenda.

However, it should be noted that the pursuit of such developments does pose fundamental problems in terms of calling forth innovations in accounting theory and practice.

As the chequered history of past attempts to develop social accounting practice indicates, a major stumbling block is encountered in the accountant's traditional obsession with, and insistence on, objectively verifiable and largely financially based measurement techniques. Schumacher (1973, p. 38) effectively pinpoints the dangers of seeking to press what are essentially non-economic values into the framework of economic calculus:

> All it can do is lead to self deception or the deception of others; for to undertake to measure the immeasurable is absurd and constitutes but an elaborate method of moving from pre-conceived notions to foregone conclusions; all one has to do to obtain the desired results is to impute suitable values to the immeasurable costs and benefits. The logical absurdity, however, is not the greatest fault of the undertaking: what is worse, and destructive of civilization, is the pretence that everything has a price or, in other words, that money is the highest of all values.

Essentially, then, for accounting to make a real contribution to the green debate, what is called for is nothing less than a fundamental re-examination of the marginalist and neo-classical economic underpinnings of the accounting craft whereby accounting can only operate when prices are generated by transfer of property rights in the market place. Whereas a financial figure can, of course, be placed upon items such as clean-up costs, it is not possible to price many elements of natural capital (air, clean water, the ozone layer, etc.) which are hence ignored by the accounting system and therefore treated largely as free goods, with consequences which are becoming ever more apparent.

In addition to moving away from an exclusive emphasis on financially quantified 'bottom line' performance measures, the necessity for which arises from the recognition that not all 'values' are quantifiable, other aspects of the traditional accounting framework stand in need of basic change in order to incorporate a green dimension. For example:

- the emphasis on single time period, historic performance measurement, and
- the concentration on the economic entity as the focus of reporting.

The point here, of course, is that ecological issues are long term in nature and particularly affect future generations whilst environmental impacts are 'externalities' as far as the entity itself is concerned.[9]

Clearly, for such a fundamental rethink to take place green concerns must be incorporated centrally within the accounting standard setting process, rather than being treated as merely peripheral. The recently outlined work programme for the newly constituted Accounting Standards Board suggests this is unlikely to occur. Indeed, experience with *The Corporate Report*, where the accounting profession saw fit to inject little urgency into the debate following publication of that document and broadly adopted a posture of sitting on the fence, gives cause for little optimism that things will be different this time round. Indeed, Jeuda (1980) has expressed the view that the real problem with *The Corporate Report* as far as the profession was concerned lay in its espousal of the concept of public accountability and promotion of the needs of other user groups in addition to capital providers. Further evidence that the traditional concern of the accountant with the needs of financial capital may prove a major stumbling block in terms of incorporating a green dimension into accounting practice is provided by Jones (1990). A programme of interviews he conducted with fifty-seven accountants working in six large manufacturing and merchanting firms elicited the overwhelming response that, in their view, profit is the prime, if not only, goal of business and social responsibilities are deserving of a very low level of priority!

In sum, one is perhaps left with the somewhat depressing conclusion that both the nature of the accounting craft and the world view exhibited by its practitioners represent major, or even overwhelming, obstacles on the road to the development of 'green' accounting. Whether an appreciation of the very real contribution accountants can make to the greening of business will be sufficient to overcome these obstacles shall doubtless become apparent in the fullness of time.

Notes

1. For further details of initiatives undertaken by the major professional accounting bodies see Owen (1991) Chapter 4.
2. For a fuller elaboration of the following argument see Harte, Lewis and Owen (1991).
3. A similar observation may be made concerning radical critique from a Marxist perspective (see, for example, Tinker, 1985).
4. The following material draws heavily from Gray's seminal analysis to which the reader is particularly referred.
5. A closely related concept being adopted by HM Inspectorate of Pollution is the 'best practicable environment option' (BPEO), introduced in the twelfth report of the Royal Commission on Environmental Pollution (1988). BPEO is defined as that which provides the most benefit or least damage to the environment as a whole at an acceptable cost in both the short and long term.
6. The reader is particularly referred to Harte and Owen (1991) for more detailed coverage of the issues raised in this section.
7. Major categories of information disclosure considered under the heading 'social' are: human resources; fair business practices; energy; community involvement; product related; and environmental.
8. Twenty-four companies were drawn from a list provided by ethical unit trusts in the context of another study concerned with the use made of published annual reports by ethical investors (Harte *et al.*, 1991). A further six companies known as being innovators in green reporting practice on the basis of previous research were added to complete the sample.
9. See Cowe (1991) for a further discussion of these issues.

References

Accounting Standards Steering Committee (1975) *The Corporate Report*, ASSC, London.

Adams, R. (1991) Why is the environmental debate of interest to accountants and accountancy bodies? in D. Owen (ed.) *op. cit.*

Blaza, A. (1991) Environmental reporting: a view from the CBI, in D. Owen (ed.) *op. cit.*

Cooper, D. J. and Hopper, T. M. (1988) (eds.), *Debating Coal Closures: Economic Calculation in the Coal Dispute 1984–85*, Cambridge University Press.

Cooper, D. J. and Sherer, M. J. (1984), The value of corporate accounting reports: arguments for a political economy of accounting, *Accounting Organizations and Society*, pp. 207–32.

Cooper, M. and Filsner, G. (1991) The environment: a question of profit – the ordinary investor and environmental issues in accounting, in D. Owen. (ed.) *op. cit.*

Coopers and Lybrand Deloitte (1990) *Environment and the Finance Function: A Survey of Finance Directors*, Coopers and Lybrand Deloitte, London.

Cowe, R. (1991) Green issues and the investor: inadequacies of current reporting practice and some suggestions for change, in D. Owen (ed.) *op. cit.*

Daley, H. E. and Cobb, J. G. Jnr. (1990) *For the Common Good: Redirecting the Economy Towards Community, the Environment and a Sustainable Future*, Greenprint, London.

Dobson, A. (1990) *Green Political Thought*, Unwin Hyman, London.

Elkington (1990) *The Environmental Audit: A Green Filter for Company Policies, Plants, Processes and Products*, Sustainability/World Wide Fund For Nature, London.

Frankel, B. (1987) *The Post Industrial Utopians*, Polity Press, Cambridge.

Geddes, M. (1991) The social audit movement, in D. Owen (ed.) *op. cit.*

Gorz, A. (1989) *Critique of Economic Reason*, Verso, London.

Grant, G. (1990) Environmental audit: cradle to grave, *Green Magazine*, Summer, pp. 53–7.

Gray, R. H. (1990a) *The Greening of Accountancy: The Profession After Pearce*, Chartered Association of Certified Accountants, London.

Gray, R. H. (1990b) Corporate social reporting by UK companies: a cross-sectional and longitudinal study. Paper presented to the annual conference of the British Accounting Association, University of Dundee, April (1987).

Gray, R. H. Owen, D. and Maunders, K. (1987) *Corporate Social Reporting: Accounting and Accountability*, Prentice Hall, London.

Guthrie, J. and Parker, L. D. (1989) Corporate social disclosure practice: a comparative international analysis, *Advances in Public Interest Accounting*, Vol. 3, pp. 67–93.

Hackett, P. (1991) Developing a trade union charter for the environment, in D. Owen (ed.) *op. cit.*

Harte, G. and Owen, D. (1991) Current trends in the reporting of green issues in the annual reports of United Kingdom companies, in D. Owen (ed.) *op. cit.*

Harte, G., Lewis, L. and Owen, D. (1991) Ethical investment and the corporate reporting function, *Critical Perspectives on Accounting*, Vol. 2, No. 3, pp. 227–53.

Hopwood, A. (1984) Accounting and the pursuit of efficiency, in A. Hopwood and C. Tomkins (eds.) *Issues in Public Sector Accounting*, Philip Allan, Oxford.

Institute of Business Ethics (1990) *Ethics, Environment and the Company: A Guide to Effective Action*, IBE, London (1989)

International Chamber of Commerce (1989) *Environmental Auditing*, Paris.

Jenkins, B. G. (1990) Environmental audit: an auditor's perspective. Text of a talk given at Glaziers Hall, London, 27 March 1990; Coopers and Lybrand Deloitte, London.

Jeuda, B. (1980) Deserving a better fate – the corporate report, *Accountancy*, February, pp. 76–8.

Jones, C. (1990) Corporate social accounting and the capitalist enterprise, in D. J. Cooper and T. M. Hopper (eds.) *Critical Accounts*, Macmillan, London.

Keynes, J. M. (1933) National self sufficiency, *Yale Law Review*, Vol. 22, pp. 755–63.

Lickiss, M. (1991) Measuring up to the environmental challenge, *Accountancy*, January p. 6.
Maxwell, S. (1990) The rise of the environmental audit, *Accountancy*, June, pp. 70–2.
Owen, D. (1991) (ed.) *Green Reporting: Accountancy and the Challenge of the Nineties*, Chapman and Hall, London.
Pearce, D., Markandya, A. and Barbier, E. B. (1989) *Blueprint For a Green Economy*, Earthscan Publications, London.
Pepper, D. (1984) *The Roots of Modern Environmentalism*, Croom Helm, Beckenham.
Porritt, J. (1984) *Seeing Green: The Politics of Ecology Explained*, Basil Blackwell, Oxford.
Puxty, A. G. (1986) Social accounting as immanent legitimation: a critique of technist ideology, *Advances in Public Interest Accounting*, Vol. 2 pp. 95–112.
Ryle, M. (1988) *Ecology and Socialism*, Radius, London.
Schumacher, E. F. (1973) *Small is Beautiful*, Blond and Briggs, London.
Seabrook, J. (1990) *The Myth of the Market: Promises and Illusions*, Green Books, Bideford.
Tinker, A. M. (1985) *Paper Prophets: A Social Critique of Accounting*, Holt Saunders, Eastbourne.
Tinker, A. M., Lehman, C. and Neimark, M. (1991) Falling down the hole in the middle of the road: political quietism in corporate social reporting, *Accounting, Auditing and Accountability Journal*, Vol. 4, No. 2, pp. 28–54.
Tonkin, D. J. (1991) Environmental protection statements, in D. J. Tonkin and L. C. L. Skerratt (eds.) *Financial Reporting 1990–91: A Survey of UK Reporting Practice*, The Institute of Chartered Accountants in England and Wales, London.
Touche Ross Management Consultants (1990) *Head in the Clouds or Head in the Sands? UK Managers' Attitudes to Environmental Issues – A Survey*, Touche Ross, London.
World Commission on Environment and Development (1987) *Our Common Future*, Oxford University Press.

6

A LEGAL ENVIRONMENT FOR SUSTAINABLE DEVELOPMENT

Neil Hawke

What is 'Sustainable' Development?

Approaching the term 'sustainable development' one has a nasty suspicion that it can mean almost anything one wishes. Certainly from the lawyer's point of view it represents, at most, a crude marker indicating that man's developmental activities require some environmental justification (Hawke, 1991). At the root of that justification, one suspects, is the deceptively simple idea that the earth's natural resources (in their widest sense) cannot be utilized without limit. Decision-making by individuals, companies, governments and regulatory agencies (to name but four) must be subject to the environmental discipline. Decisions can no longer be self-centred, they have to take account of a growing awareness of the need for development to be sustainable by reference to finite environmental resources.

A Meeting of the Ways

Principles of sustainable development seem to fall into two categories from a lawyer's point of view, according to whether they receive explicit recognition in law, or not. This categorization probably raises a further issue focusing on the variables which may determine whether any one concept of sustainable development should receive legal recognition. If so, what variables will determine the nature and scope of that recognition? Some possible answers to these questions will be examined later. For the moment, though, the foregoing reference to a meeting of the ways emphasizes that, increasingly, sustainable development forces lawyers and the world of business into a dialogue. Even if the word dialogue is misplaced it is frequently the case that major planning inquiries force the proponents, planning authority, Secretary of State and objectors into a lively debate on the issues surrounding sustainable development. If this is one of the most visible fora for debate and argument it is often accused (perhaps rightly) of being

- site specific
- unable to debate the merits of government policy.

One of the best known illustrations of the constraints affecting public inquiries relates to motorway inquiries where the event is often characterized as involving 'salami politics' in so far as discrete inquiries may be held in respect of separate sections of the proposed motorway rather than the *whole* enterprise.

Some Early Examples: BPM

From the lawyer's point of view one of the most visibly familiar and practical manifestations of sustainable development comes from the requirement to adhere to 'best practicable means' (BPM). BPM makes a reappearance in the Environmental Protection Act 1990 where (in Part III) the Act deals with statutory nuisances.[1] The definition of BPM follows the original definition that remains in the Control of Pollution Act 1974 in referring to a variety of factors.[2]

More particularly, BPM is to be interpreted by reference to an expanded meaning for the word 'practicable' which is stated to mean

> reasonably practicable having regard among other things to local conditions and circumstances, to the current state of technical knowledge and to the financial implications.[3]

While other factors are to be included in the interpretation of BPM, the foregoing extract sits at the centre of any application of BPM. Perhaps significantly BPM provides a defence to certain criminal proceedings as well as grounds of appeal, for example in proceedings for statutory nuisances.[4] The main beneficiary in both cases tends to be anyone engaged in activities on industrial, trade or business premises.

A better known application of BPM comes in the Health and Safety at Work Act 1974: section 5 (for example) requires a person controlling certain categories of premises to use BPM as a means of preventing emissions to the atmosphere. In this case the Act only goes so far as to say that 'means' includes

> a reference to the manner in which the plant provided for those purposes is used and to the supervision of any operation involving the emission of the substances[5]

Elsewhere in the Act of 1974 there are references to legal duties that are to be performed if it is 'practicable' or 'reasonably practicable'. The relationship between these terms and BPM is seen in the foregoing references to BPM definition in the Environmental Protection Act. Any BPM must be 'practicable' or 'reasonably practicable'. There is no doubt that that Act's prescription is potentially very helpful in its detailed descriptions of the variables, and perhaps more helpful than the courts' attempts at definition:

> if a precaution is practicable, it must be taken unless, in the whole circumstances, that would be unreasonable (Lord Reid in *Marshall* v *Gotham*)[6]

Hidden Agendas?

While there are explicit references to sustainable development in the above statutes and elsewhere, in the Clean Air Acts, for example, it can be said that these references are merely first steps in a very complex attempt to provide a balance between development and environmental protection. Nevertheless, the so-called 'first steps' may be significant in raising the profile of sustainable development as an express element in decision-making or in enforcement. Other techniques may follow in aiding enforcement of such standards, for example. This is seen in the Health and Safety at Work Act which stipulates that in any proceedings for a breach of duty, the onus of proof is reversed *and is on the accused* who is then obliged to prove that it was not practicable or not reasonably practicable to perform the duty in question.[7] On occasions when evidence is difficult to find this is a considerable aid to enforcement.

The lawyer's preoccupation with certainty means that there is a comforting reassurance in dealing with the explicit references to sustainable development principles. However, it is clear that there is a much bigger hidden agenda continually able to influence decision-making affecting the environment. There is no better example of this than 'best practicable environmental option' (BPEO), a term that (again) is heavily influenced by measures that seek to determine whether a proposal is 'practicable'. The Royal Commission on Environmental Pollution reported on BPEO in 1988[8] and observed that

> The indiscriminate use of the term to describe almost any course of action which takes some account of environmental factors can only undermine the underlying principles on which BPEO is based.[9]

Thereafter the Royal Commission sets out a process for determining a BPEO. Such an approach is very laudable in view of the concern expressed in the extract set out above. Unfortunately we are left with a bewildering variety of approaches which are often incapable of being tested through planning inquiries, for example. Not only is there a bewildering variety of approaches. BPEO is a tool that has been used for a wide variety of purposes from policy-making to hard decision-making. There may be some reassurance on occasions when BPEO criteria are crystallized in relevant Codes of Practice: Waste Management Paper no.26 on the land-filling of wastes is a significant example.[10] In the same context, however, the Environmental Protection Act indicates in s.35(8) that

> It shall be the duty of waste regulation authorities to have regard to any guidance issued to them by the Secretary of State with respect to the discharge of their functions in relation to licences.

Despite the tenor of the Waste Management Paper mentioned previously there is really no reason why the powers in s.35(8) should not be used for the purpose of imposing the Secretary of State's own agenda for BPEO in decision-making on waste management licence applications. For the moment though BPEO does receive one explicit reference in the Environmental Protection Act's reference to Integrated Pollution Control (Part I). There, the authorization process by which selected industrial processes may be subject to controls imposed by Her Majesty's Inspectorate of Pollution states that the BATNEEC criteria should have regard to BPEO in relation to the substances that will be released to the environmental media.

BPEO Unleashed

We can now see a growing number of instances where BPEO is influential in a formal context of statutory recognition. Two very significant examples are:

- Integrated Pollution Control (IPC) under Part I of the Environmental Protection Act
- Environmental Assessment (EA) under the Town and Country Planning (Assessment of Environmental Effects) Regulations 1988.

The first of these examples involves enforcement by Her Majesty's Inspectorate of Pollution of multi-media sources of pollution. Any manufacturing operation included that generates polluting substances to land, water and air will be subject to unified regulation according to BATNEEC: best available techniques not entailing excessive cost. This concept is one that has been recognized in EC Directives on the Environment for some time: see, for example, the Framework Directive on Air Quality[11] and the Dangerous Substances in Water Directives.[12] The intention is that licensing will be undertaken by reference to certain objectives which include BATNEEC. The requirement of the Environmental Protection Act is that BATNEEC will be used to minimize environmental pollution, having regard to the best practical environmental option for substances to be released. The consent granted here is intended to be a complete statement of the technology to be used on the site. Underlying the concept of BATNEEC is the need to reduce emissions and discharges to their lowest practicable level in the case of the particular process. The individual elements of BATNEEC can be summarized as follows:

- *Techniques*. These refer to the process (including its concepts and design) and its operation as well as its components and their interrelationship, extending to staffing, training, work methods and supervision.

- *Available*. The technology should be available to an operator of the type of process in question. Even though that technology is not generally in use it should be generally available even though it may be available solely through a monopoly supplier.

- *Best*. The technology must be seen in relation to the process of prevention, minimization and neutralization as being the most effective technology.

- *Not entailing excessive cost.* This factor is a potentially influential variable that can impact on BAT. In simple terms, achievement of BAT may be excessively costly by reference to any environmental benefit. In these circumstances BAT would probably be qualified in the face of these economic forces.

In the second of the examples relating to EA the system of planning control under the Town and Country Planning Act 1990 plays host to a requirement that certain types of development should be submitted to an environmental assessment before a decision is made on a planning application. More particularly the requirement, whose source is EC Directive 85/337, is either automatic for some developments (power stations, for example) or dependent on a judgement about any significant effects on the environment. BPEO seems to be implicit in those parts of the Regulations dealing with the content of the EA. In particular there is a need to identify (among other things):

- likely significant effects, direct and indirect, on the environment of the development
- (if significant adverse effects are identified) a description of the measures envisaged in order to avoid, reduce or remedy those effects
- (in outline) the main alternatives studied by the applicant
- any difficulties, such as technical deficiencies or lack of know-how.[13]

ALARP and ALARA

Another approach to the principle of sustainability comes from the area of licensing governed by the Nuclear Installations Act 1965. The Health and Safety Executive is advised here by HM Nuclear Installations Inspectorate. In assessing safety provision reference is made to the design of the plant and its ability to meet the lowest reasonably achievable exposure of workers and the public to radiation. ALARA ('as low as reasonably achievable') balances the benefit of radiation dose reduction together with its health and social implications, with the overall cost of resources and any additional harm involved in achieving the reduction. The National Radiological Protection Board's view is that ALARA does not require everything possible, or technically feasible, to be done to reduce doses without regard to cost or the magnitude of benefit from dose reduction. This standard is enforceable in law through the medium of the Ionizing Radiations Regulations 1985.

By way of contrast HM Inspectorate operates a set of Safety Assessment Principles in dealing with a licence application for a nuclear power station. In this case the principle used is ALARP: 'as low as reasonably practicable'.

Although the two principles, ALARA and ALARP, are difficult if not impossible to distinguish in effect, they are applied and enforced by different agencies. Consequently there is a danger of inconsistent interpretation, a danger adverted to by Sir Frank Layfield at the recent Sizewell Inquiry. Despite similarities it is often worth emphasizing that ALARP has its roots firmly in the BPM/BPEO tradition and the common law. Consequently the focus of responsibility is firmly on the licensee. Nevertheless one of the recommendations from the Sizewell Report is that guidance is needed on the application of ALARP.

The Polluter Pays Principle

This principle (PPP) was defined some years ago by OECD as follows:

> the polluter should bear the expense of carrying out the ... measures decided by public authorities to ensure that the environment is in an acceptable state.[14]

PPP does focus attention on the growing interest in economic instruments for environmental protection, dealt with later. Indeed, the House of Lords Select Committee on the European Communities, in its 1983 Report on PPP,[15] bracketed such instruments with BPEO in referring to the

> two mechanisms which could maintain pressure to improve environmental quality.[16]

The Select Committee also recognized advocacy of

> the belief that control by the automatic operation of market forces is more efficient and more flexible than control by regulation.[17]

Elsewhere, OECD in its recent publication on *Economic Instruments for Environmental Protection*[18] recognizes that

> economic instruments rarely act as a substitute for direct regulations. Instead the choice is among various combinations of direct regulations and economic instruments.

Ultimately the Select Committee confirmed that PPP could not be regarded as a scientific principle capable of precise definition. Indeed,

> a mix of different practices among member states in application of PPP [is] likely to entail some distortions in trade.[19]

This latter point can be illustrated in a variety of ways. One notable example recently has been the so-called Danish Bottles case, *Commission* v *Denmark* (1988)[20] where bottle recycling in that country was challenged successfully in the European Court of Justice as a breach of Article 30 of the Treaty of Rome dealing with the free movement of goods between member states. The restriction on beverages that could be marketed in non-approved containers was seen as being in breach of Article 30. The decision of the court is significant in so far as it does not indicate the extent to which any member state's requirements for environmental protection can be subject to control and restriction under EC law. In the same vein a recent report for the Department of the Environment by Environmental Resources Ltd[21] shows that Germany is the only country reviewed in the survey of *Market Mechanisms* that had introduced pollution charges with a primary incentive objective. It also observes that the volume and value of subsidies available for environmental investment exceeds those available elsewhere in Europe. Not surprisingly, the Report goes on to conclude that subsidies are

> contrary to the strictest interpretation of the Polluter Pays Principle. When general taxation is used to finance subsidies the polluters' costs are being borne by the public at large, which may be considered inequitable.[22]

Economic Instruments

Considerable debate has been triggered recently on this subject, particularly through the work of Professor David Pearce for the Department of the Environment. The White Paper recently published confirms the need for a foundation of regulatory, administrative controls. However, the White Paper goes on to list the following items:[23]

- market-based instruments
- cost recovery charging
- levies
- incentive charging
- product charging
- tradable quotas
- legally enforceable financial liabilities
- subsidies and schemes of public compensation

From a personal point of view the list shows a very significant relationship between the lawyer and the environmental economist. Both probably need to collaborate far more closely as we see a diversification of the environmental agenda.

Conclusions

There is no doubt about the dimensions of the picture painted in this chapter: the need for more effective and sophisticated environmental disciplines demands a set of principles for sustainable development. That quest must involve a dovetailing of regulatory techniques through the law together with economic and related instruments. All too often the law will provide a delivery vehicle as well as a facility for underpinning responsibilities through well-accepted approaches such as the imposition of strict liability.

Somewhere in all of this, principles of sustainable development have to be recognized. From the perspective of the law, these principles may (as seen previously) be accommodated as part of the law itself, either in statute or in common law principles developed by the courts, or be 'free-standing' criteria. As free-standing criteria (albeit contained in a Code of Practice) the principles are used for a variety of purposes: two plain examples are in the licensing process and in many policy-based decisions taken by government.

The fundamental problems affecting these principles of sustainable development where they are not formally contained and recognized in law can be enumerated as follows:

- uncertainty on the question whether issues affecting sustainable development will be part of the relevant agenda for environmental action
- consistency in the application of sustainable development principles across a variety of areas of decision-making or other action
- absence of any coherent guidance about the relevance of some of the elements and the weight to be given to them
- selective reliance on the principles as between the creation of environmental policy and its implementation
- inability to challenge the relevant criteria where, for example, BPEO is used in the relatively closed world of policy-making compared with the public participation that may be allowed at the decision-making level.

Throughout any examination of sustainable development criteria there is a preoccupation with relative variables. In determining what is 'reasonably practicable', for example, it will often be clear that industry cannot be expected to adhere to 'outlandish' requirements for environmental protection. In one way or another the law and its framework can be expected to perpetuate this concept. Although BPM and BATNEEC, for example, may be able to give some insulation from more extreme financial commitments in the pursuit of better environmental protection, they hardly force the pace. If the government and/or the EC choose to force the pace a more potent, market-based approach is needed in which legal regulation is more explicitly linked with economic instruments. The classic example is incentive charging pitched at an appropriate level!

At the outset, in referring to 'a meeting of the ways' reference was made to legal recognition of sustainable development concepts. The first variable determining legal recognition has been through the courts' preoccupation with a distribution of liabilities in the law of tort where standards of reasonableness have been the currency of the law for so long. Developing legislative policy has seen adoption – and adaptation – of these standards, particularly in the use of BPM in various statutory provisions. The important variable here is again the need to produce rules directing and distributing liabilities and responsibilities. This task is now further refined and developed in the context of EC Law although, perhaps fortunately, the roots of the development are still recognizable to the English lawyer!

Notes

1. Section 79.
2. Section 72.
3. Ibid. note 1.

4. Environmental Protection Act 1990, Part III.
5. Section 5(2).
6. [1954] 1 All E.R. 937 at 942.
7. Section 40.
8. 12th Report.
9. Ibid. para. 1.6 at p. 2.
10. HMSO 1986.
11. 84/360.
12. 76/464.
13. Schedule 3.
14. Recommendation of the OECD Council (26 May 1972), para. 4.
15. 10th Report, Session 1982–83.
16. Ibid. at para. 14.
17. Ibid. at para. 17.
18. OECD, Paris 1989 at p. 120.
19. Ibid. note 16 at para. 26.
20. Case 302/86.
21. January 1990.
22. Ibid. p. 61.
23. Cm. 1200, HMSO 1990 at Annex A.

References

Hawke, N. (1991) *Nature Conservation: The Legal Framework and Sustainable Development*. Leicester Polytechnic Law School Monographs.

7

EXPLORING THE GREEN SELL: Marketing Implications of the Environmental Movement

Jo McCloskey, Denis Smith and Bob Graves

Introduction

Along with decaffeinated coffee, safe sex and alcohol-free beer, the environment emerged in the late 1980s as a potent marketing tool for retailers and manufacturers alike. The concept of being 'environmentally friendly' has captured the imagination of marketers and product managers, many of whom now claim that their products have a long and distinguished pedigree in terms of environmental impact. However, environmental concern is a nebulous concept which cannot easily be fully translated into details of product performance. Many 'environmentally friendly' claims cannot be adequately tested in terms of environmental degradation due to the complex trans-scientific nature of the pollution problem. The purpose of this chapter is to explore some of these issues by reference to a number of sectors and, in the process, relate the impact of environmentalism to mainstream marketing theory.

Marketing and Environmentalism

Whilst public concern about environmental quality is not new, the use of such anxieties by marketers to sell products and services has a relatively short history. Prior to the recent wave of concern, the benefits projected as being associated with products tended to focus on the wants and needs of the buyer or the immediate family/reference groups which they represented. However, many of the 'new wave' of environmentally friendly products have been marketed in such a way as to engender within the consumer concern for wider environmental issues.

To date the main expression of concern has been manifested in the process of waste reduction through product purchasing and recycling potential. In a survey for the *Wall Street Journal* in 1991, some 45 per cent of Americans questioned felt that they should do something to improve environmental quality via their purchasing patterns (Simons, 1991). Whilst this contrasts with a staggering 82 per cent in Germany (ibid.) it still represents an important shift in the buying behaviour of North Americans, with as many as 90 per cent of a sample interviewed by Proctor and Gamble indicating that they had environmental leanings (Alvord, 1991). The problem of packaging and solid waste disposal in the USA is indicated by the arisings of some 806 million tonnes per year of waste by comparison to 354 million tonnes in Japan (Simons, 1991). Of this total, the USA recycled 10 per cent, incinerated 10 per cent and landfilled the remainder. In contrast, Japan recycled 50 per cent, sent 34 per cent for incineration and only landfilled 16 per cent (ibid.). It is obvious from the scale of the problem that steps needed to be taken to arrest the growing landfill burden in the USA, especially by industries such as cosmetics where critics have suggested that the bulk of the product is packaging (Roddick, 1991). The potential exists, therefore, for companies to make considerable reductions in the waste potential associated with their products and, given increasing societal concerns, derive marketing advantages in the process.

It is possible to identify five principle elements of a generic green marketing strategy. The first element concerns those direct benefits that accrue to consumers in terms of environmental quality. New 'ozone friendly' refrigerators provide an illustration of this, where the proposed refrigerants would be less damaging to the environment than the CFCs currently used. The potential benefits in developing such a refrigerant are considerable as the market is estimated to be currently worth approximately $5 billion (Evans, 1990). Obviously any product which reduces personal risk or impacts upon the individual's immediate environment will have a certain cachet for the consumer. Products which are proven to cause less pollution because of their physical or chemical properties or because they are intrinsically safer also have obvious marketing potential.

Some products have only indirect benefits in terms of environmental quality and these have to be dealt with differently by marketers. Products such as concentrated washing powders, which claim environmental benefits because less powder is used, provide a case in point. Such claims could be regarded as spurious if one considers that manufacturers are only claiming alternative characteristics for existing ingredients and technologies instead of looking for other, less harmful, substitutes. Indeed it could be argued that these manufacturers are more concerned with stressing their corporate green image in order to give their products legitimacy within the framework of environmental concerns. Taking the example of concentrated washing powders further, because a powder is more concentrated does not mean that its environmental impacts will be less. Indeed, it could be argued that certain powders may cause greater impact, because of the presence of enzymes, than more traditional and less concentrated powders.

A third element takes the form of attempts to link the product to a respectable environmental organization, such as Friends of the Earth, in order to give the product greater legitimacy. The marketing of a cause is not a new phenomenon but organizations have been quick to identify that advertising such a link can provide a greater degree of credibility for their products and enhance their corporate image. Perhaps the best example of an effective public relations-marketing strategy has been that employed by British Nuclear Fuels Limited (BNFL) at its Sellafield site in Cumbria. The company has succeeded in turning around the site's image from being a heavily polluted industrial complex to a plant with an almost 'theme park' image. The visitors' centre combined with an aggressive TV advertising campaign have transformed the site into perhaps the most popular single site attraction in the English Lake District.

The fourth element within a generic marketing strategy concerns the promotion of a corporate image, such as BP's use of a logo which emphasized the 'green' aspects of the company's business. However, as BP found to their cost, attempts to market a green image will eventually require companies to live up to the marketing hype.

Finally, the period over which the company and/or the product is deemed to have been environmentally friendly is also of importance. A number of products are currently being promoted on the basis of having always been kind to the environment. These include sanitary products such as Tampax which claim a long history of being 'kind' to the environment. The rationale behind such a move is to reassure the consumers of both the company's and product's 'proven' worth in this respect.

The various elements of green marketing, listed above, raise a number of serious questions that need to be addressed by marketers. The first relates to marketing's basic function to link production and consumption within the context of societal demands. Marketing should allow the business enterprise to supply products which the market wants. The question that needs to be addressed is whether marketing is shaping those desires rather than responding to them. This takes us into the second issue, namely, are industries using socially responsible marketing practices with regard to environmental issues. The remainder of this chapter seeks to address these two issues and suggests the implications that they hold for future marketing strategies across a range of company types and activities.

The Customer's Interest

The marketing concept defines marketing as a philosophy which states that a firm's activities should be based on satisfying customers' needs and wants in selected target markets (Gronroos, 1990). This market-orientated view, however, must consider that in an age of environmental deterioration and resource shortages, an organization's tasks should begin by determining the needs, wants and interests of target markets and to deliver products or services that satisfy those desires not only effectively and efficiently, but in a way that preserves and enhances the consumer's and society's interests in the long term (Kotler, 1990).

Originally, company decisions on marketing strategies were based on costs and profit forecasts. However, companies today must realize that satisfying customer demands goes further than supplying the market with a functioning product. Consumers now want to ensure that products are safe, both to themselves and to their environment. These concerns have led to a powerful movement within society which seeks to improve the rights and powers of buyers in relation to sellers. Many members of the public believe that the balance of power rests with the manufacturers and retailers because consumers have too little information, education and protection to make informed purchase decisions when confronted with sophisticated marketing techniques. Pressure from consumers has resulted in moves to establish high standards of quality, via a British Standard, wherein the corporation would be liable for a failure in a product which arises from a design or manufacturing shortcoming.

This mistrust on the part of the consumer seriously conflicts with the general marketing tenet that the customer is the core to any marketing strategy and that all marketing decisions should be focused on customer needs and wants. A major reason for this anomaly centres around the conventional application of the marketing concept to practical business situations. A successful marketing strategy is one that is seen to achieve a mix of marketing variables which yields the greatest profit. The limitations of this approach have become increasingly obvious but it is still widely used as a basis for marketing decision-making, despite the fact that it takes no direct account of customers' wishes (Gronroos, 1990). This creates a fundamental flaw in the application of the marketing concept which continually states that an organization bases its activities on satisfying customer needs and wants. One can argue that companies undertake marketing research to identify what the market wants, but the findings of such surveys are merely made to fit into this marketing mix model, thus implying that companies are still somewhat process and product driven rather than market driven.

One of the major impacts of continuous consumer research, and the publicity given to its findings, is that the consumer is becoming more informed about the offerings in the market place. They are now more educated on aspects of the quality and make-up of the products, and of the long-term effects that the processes of production will have on their environment. Consumers have been expressing these concerns since the 1960s but now they are beginning to demand that they be taken seriously by manufacturers and retailers alike. In order to place environmental considerations at the centre of consumer production processes, the public have begun to use their purchasing power more discerningly in an attempt to exercise more influence over issues that affect the quality of their lives.

It could be argued that manufacturers and retailers have been slow to find ways to improve their products and processes of production in order to meet the changing shifts in attitudes of their target markets. Their response to green market demands has often been one of minor product or package modification and has varied from merely labelling a product as being 'environmentally friendly' or reproducing the product in a more concentrated form, thereby reducing packaging and waste. This type of strategy is superficial at best and implies that established brand manufacturers appear to be sticking to marketing strategies based on the principle that a good name sells and any deviation from the tried and tested product offering may be counter-productive and damaging to corporate image and brand loyalty. This stance, whilst effective in the mass-producing and mass-consuming 1960s and 1970s, is no longer an accurate interpretation of modern consumer behaviour set within the context of sustainable development. On the basis of market research carried out in the USA, Proctor and Gamble found that 62 per cent of consumers wanted and read product information regarding environmental performance and 41 per cent said that it influenced their buying behaviour (Alvord, 1991). The USA is much further advanced in terms of product labelling than the UK and this is reflected in the strategies adopted by US companies. This experience has suggested that positive benefits arise from product labelling as it allows companies to reach 100 per cent of their product users, helps the consumer gain a better understanding of the nature of environmental problems and allows them to make informed choices on environmental performance with respect to other, competing products (Alvord, 1991). The implications here with regard to corporate responsibility are considerable and these will be returned to later in this chapter.

One of the major weaknesses of the strategy of packaging reduction is that it overlooks the principle that consumers also buy product benefits. Whilst reductions in packaging are to be encouraged in an attempt to reduce the solid waste burden, manufacturers must also strive to ensure that the remaining packaging is biodegradable and that the materials that are being sold are not themselves harmful to the environment. For example, washing powders and liquids which come in reduced packaging may still be contained in non-biodegradable material and may cause environmental degradation through the presence of enzymes as cleaning agents. Similarly, claims of recyclability are void unless the infrastructure exists to facilitate the recycling process at the point of sale/use. In the USA, for example, Rhode Island requires that product statements concerning recycling can only be made if there is a proven infrastructure in place (Alvord, 1991); and this is in keeping with the EPA's hierarchy of waste reduction policies which places source reduction at the head of the list (Barnett, 1991). Consequently, product benefits must be both *proven* and *realistic* in order to serve any useful purpose in improving environmental quality. It is also obvious that there is a necessity for corporate and governmental cooperation to establish the infrastructures needed to implement recycling programmes. In terms of product performance there is a necessity for legislation which demands product labelling to facilitate *effective* consumer choice amongst competing brands. Two of the product benefits that are currently being sought by environmentalists are: a reduction in personal risk and, at a more general level, effective pollution prevention.

Personal Risk

Many products are now targeted at that sector of the population who want to minimize the risks to their health associated with certain products. The markets for additive-free food products, decaffeinated coffee, alcohol-free wines and beers and bottled water can all be related to the consumer's desire for a lower level of risk within their life-style. Similarly, the whole issue of product liability can be traced to concerns expressed over deleterious effects associated with product failure.

Recent publicity concerning contamination in pâtés, soft cheeses, eggs and bottled water and the continuing risk of harmful bacteria in cook-chill and frozen foods have all heightened public concern over quality control. Food retailers must avoid any negative associations connected with the food production chain if they are to retain the public's confidence in their products. People want to be certain that the food they buy is of a high standard and will not have any long-term detrimental effects on their well-being. However, most manufacturers and retailers have responded to this public concern by introducing alternative packaging which contains an 'additive-free' label and instructions for storage. As yet, there is no prescribed limit to the presence of bacteria in fresh food and the new regulations in the recent Food Safety Act do not adequately protect the consumer from the risk of such contamination. This has created an atmosphere where manufacturers can afford to be conservative in their approach, and the government appears reluctant to establish enforceable guidelines on what constitutes satisfactory food (McMurdo, 1990). In a related area, that of product tampering, companies were much quicker to respond to public concerns over third-party interference with such products as baby food, medicines and certain bottled goods. The industry responded with tamper-proof containers and used the change in the packaging format to change their marketing strategy.

Some companies are responding to such public concern and are engaged in real reforms such as looking for less harmful product ingredients and revising production processes to conserve energy and eliminate waste. Others are, however, merely cashing in through green advertising and marketing campaigns in an attempt to demonstrate to the public that they are responding to customer demands. Supermarket shelves have subsequently blossomed with 'free-range' animal products, 'additive-free' foods, 'organically grown' fruits and vegetables, 'CFC-free' aerosols and biodegradable toiletries. Such strategies are often driven by notions of economic and legal responsibility rather than by the much broader questions of ethical and discretionary responsibilities which characterize the more enlightened corporations.

The boom in environmentally friendly products and packaging has generally been consumer-led, with a strong influence from mainland Europe, and the multiple retail chains have been quicker than the branded market leaders to capitalize on this by producing their own brand of 'ecological' or 'green' products. With their new improved formulations and packaging, they have amply demonstrated that acceptable, easy alternatives to environmentally unfriendly products can meet with success. The quick response by retailers, which has been supported by smaller, lesser known manufacturers (such as Ark and Ecover), has given the consumer a wider choice in the selection of products that not only carry less personal risk but also cause less damage to the environment. The introduction of 'own label' green products has had the effect of segmenting the 'green' market into the more affluent purists, who buy branded green products, and those who buy 'own-brand' labels. This goes some way to dispelling the theory that green consciousness is a passing fad which is solely the preserve of the AB classes and demonstrates that changes have been fuelled by grass roots pressure and changing demands from a better informed public. Multiple retail chains have been forced to alter their practices to take account of, and respond to, profound shifts in consumer attitudes to consumer goods.

However, while some of these innovations can be regarded as genuine, others have been exposed as the same unfriendly product that was sold before, but in a different package or sporting an 'environmentally friendly' label. Such claims are at times spurious and are merely used to create a competitive advantage, or even confuse the consumer. It would have been inconceivable, for example, five years ago for a marketing company such as Proctor and Gamble to promote its Ariel Ultra washing powder, with an £8 million advertising campaign, claiming the twin benefit that it not only cleans but is also 'kind' to the environment. Such claims are often difficult to substantiate fully and can include waste reduction, biodegradable packaging and recyclability in addition to the environmental benefits associated with the product itself. Unfortunately, many products focus on the packaging elements rather than on their environmental performance. Perhaps one of the most cynical claims came from Higgs Furs, whose advertisement in 1989 promoted fur as 'environmentally friendly' on the grounds that it does not deplete the ozone layer or pollute the atmosphere, unlike the processes used to manufacture fake fur! The problem created by these environmental claims is that, unlike performance claims, they cannot be easily tested by the consumer. However, they are increasingly coming under scrutiny by more environmentally conscious shoppers who are seeking tangible benefits associated with product use. Similarly, retail chains can claim to be environmentally friendly by virtue of the fact that they stock a small range of green products. At no point do their own operating procedures relating to energy conservation and recycling enter into the equation. It is only the more enlightened companies, like Tesco who have taken the green challenge fully on board, which seek to change their own culture in addition to their product range.

If manufacturers and retailers were applying the basic marketing concept of responding to customer needs and wants, they ought to be seeking to use such changing consumer desires for safer and purer products as an opportunity to improve the environmental performance of both their products and operating procedures. This response to consumer demands would also strengthen and develop customer relationships whilst strengthening brand loyalty. As it is, many enlightened consumers remain sceptical of environmentally friendly claims and suspect that manufacturers and producers are using greenness as just another opportunity to increase sales or profit margins. If consumer companies continue to 'green' their images – and make a lot of noise about it in the process – whilst not actually offering any green products, they will lose credibility and eventual market share to the multiple retail chains who are in the vanguard of the green revolution.

Pollution Prevention

While consumers are concerned about the quality of products and the long-term effects that certain additives and chemicals can have on their personal well-being, there is also concern that these products and the processes by which they are manufactured are damaging the environment. This growing public concern about the environment initially posed a threat to consumer-product companies. Pollution prevention has become a major concern in recent years and the current wave of environmentally friendly products is underpinned by a desire to arrest the process of global, as well as local, environmental degradation. Any products which aid, or are perceived to aid, the consumer in this process, whilst not entailing excessive additional costs, will have a competitive advantage. Now the market is saturated with exaggerated claims of less packaging, recyclable glass, plastic or paper. However, as we argued earlier, these claims are worthless without the supporting infrastructure to ensure that recycling claims can be implemented. Similarly, consumers are continually being exposed to promotions for 'phosphate-free' detergents, biodegradable toiletries, mercury-free batteries and CFC-free aerosols. Again there is a need to ensure that the substitute components of the product do not themselves cause environmental damage.

Such green claims can, however, rebound on those companies making them, as Proctor and Gamble discovered when they launched Ariel Ultra washing powder. The product was promoted as more ecological than rival brands because it used less packaging and chemicals, but their green image became tarnished when *Today* newspaper ran a front page article pointing out that Ariel Ultra had been tested on animals. Similarly, British Petroleum's attempt to green its company logo also ran foul of the process by attracting criticism from environmental groups who pointed to the company's environmental record. By making their logo a shade greener, the company was exposed to severe criticism which actually proved counter-productive to their initial aims (Anon, 1990).

Today's consumers not only look at whether the marketing system is efficiently serving their needs and wants adequately, they are also concerned about how marketing affects their environment. Manufacturers certainly must produce offerings that satisfy consumer wants but they also have a responsibility to ensure that those goods and their production processes are not harmful to the individual or to the environment. Consumers can make specific demands for non-toxic and environmentally friendly production and disposal processes but they are not normally in a position to recommend alternative, safer processes. This must come from the producers themselves, or from government agencies, as a response to public demand. Alternative raw material substitutes and methods of production must be found in addition to packaging and product changes, if manufacturers are to be seen to be acting in a socially responsible manner; otherwise, they will not be able to sustain the public's confidence in them, their operations and products.

British manufacturers currently appear to remain lacking both in the imagination necessary to seize hold of new opportunities and in the determination to *genuinely* clean up their act rather than just clean up their image. Fear of higher costs and lower profits through a loss of competitive edge plays a large part in their reluctance to tackle environmental problems. However, firms cannot operate for long against the tide of public opinion and green concerns will have to be properly addressed if organizations want to remain in today's market place. This is particularly so if the 'level playing field' demanded by many sectors of industry happens to be configured to serve German standards. In a country where the power of the green consumer has led to significant changes in corporate behaviour, unsubstantiated claims concerning environmental quality are not easily tolerated. This climate of opinion, in which 82 per cent of consumers use environmental criteria within their buying behaviour (Simon, 1991), serves to create a powerful barrier to entry for companies who only pay lip-service to green consumerism.

Current Green Strategies

Industry has adopted several different strategies in response to environmental concerns, but to date, most of these can be regarded as fire-fighting strategies. Firms such as British Nuclear Fuels Limited, Shell and BP have relied on corporate advertising and donations to environmental causes, to sensitize the public to their point of view in an attempt to counter claims of environmental pollution that have been levelled against them. This type of strategy ignores the fact that their products/services will need to incorporate new environmentally friendly attributes and that future generations of consumers will no longer be satisfied with insubstantial promotional claims, or with strategies designed to promote the corporate image without making fundamental changes to organizational culture and practice.

Other organizations, particularly car manufacturers, have modified their products in some way in order to steal a march on imminent legislation. Catalytic converters are now standard in Audi and BMW cars and all new models of cars can run on unleaded petrol. However, lead-free petrol contains other pollutants, and catalytic converters change carbon monoxide and hydrocarbons into a combination of nitrogen, water and carbon dioxide, but carbon dioxide is one of the worst 'greenhouse gases' (Smith and Sambrook, 1990). In addition, the Californian standard for exhaust gas emissions has been questioned concerning its suitability for use in temperate Northern Europe. One of the problems that car producers face is that historically the car has been marketed as essential, desirable, and an expression of power and status. A more environmentally friendly car is simply one that is smaller with a less powerful engine, but this does not fit easily with the traditional sales pitch. Manufacturers such as Volvo, Volkswagen and BMW have also used safety as a key marketing tactic in combination with extra product features such as fitting lean-burn engines. Volvo, in particular, have sought to include safety and environmentalism within their promotional strategies and have combined these with the more traditional concepts of performance and 'status'.

One of the reasons that firms are relying on promotional rather than marketing strategies to tackle the green issue is that industry's responses to market demands are reactive and based on models of consumer behaviour that were effective in a mass-consumerist society. The tactics advocated by some of these models are not necessarily effective on today's more sophisticated consumer, set within the context of environmentally sustainable development.

Changing Patterns of Consumer Behaviour

Environmental consciousness, and the part it plays in consumer purchase decisions, demonstrates a fundamental shift in consumer behaviour. Most people, having achieved a comfortable and affluent life-style, are moving from purely material values to the more non-material aspects such as the quality of life. This change in consumer behaviour can be explained, to some extent, by looking at Maslow's 'hierarchy of needs' theory (Maslow, 1954). Abraham Maslow argued that consumers are motivated by a five-level need hierarchy, where the lower-level physiological needs (hunger, thirst) must be satisfied before the higher-level needs of self-expression are activated. It could be argued that human reactions to environmental issues fit into the upper end of the hierarchy, hence strongest concerns for environmental protection tend to centre around those countries where living standards are high, but are not so pronounced in less successful economies, such as those in the developing world and the former Eastern Communist bloc. If this explanation of the change in consumer behaviour has any validity, then environmentalism is a naturally evolved process, and not a temporary aberration. Manufacturers must therefore recognize that consumers' values are changing to reflect societal and environmental considerations rather than mass production and consumption.

This environment consciousness and green consumer behaviour will diffuse through consumer and industrial markets at an increasing rate. Industry will need to respond to this environmental issue as public awareness increases and impacts on marketing techniques. Just as marketing practices underwent a profound change with the advent of mass production in the 1960s, so too will environmentalism bring about a revolution in the provision of goods and services in the 1990s. The market has already witnessed trends towards organic food production and ethical investment policies. The consumer is also beginning to regard excessive packaging and exaggerated claims to greenery as negative selling points.

The issues involved in green marketing may eventually see consumers rejecting the throwaway attributes of products and returning to attributes such as quality, reliability, durability and safety (Borden, 1991). It is reasonable to assume that those companies who can innovate and implement products and processes that are favourable to the environment will gain a significant competitive advantage within the markets they serve. The more widely environmental issues are promoted and understood by consumers, the greater will be the effect on consumer behaviour, and the greater the resulting impact on the product-market strategies pursued by manufacturers. These issues, if not properly addressed, could have a destabilizing effect on certain industries and marketers will have to devise more ethical and proactive strategies if these industries are to survive.

Ethical Dynamics of Green Marketing

The ethical aspects of marketing have received some consideration in the literature since they were introduced on to US business curricula during the early 1970s. Within the context of the present discussion it is apparent that the marketing of a product's environmental and safety characteristics have obvious ethical implications. For example, to claim that a product has a beneficial impact on environmental quality on the basis of weak evidence is obviously morally wrong and potentially damaging to the corporate image in the long term. However, if we go beyond the obvious issue of product quality it is clear that there are other, hidden aspects of an ethical approach to marketing. Vinten (1990) suggests that broad issues of bribery, used in breaking into certain markets in the developing world, can 'lead to instability in the pricing structure' (p. 5) especially in the developing world where they have an impact on the marketing mix (see also Kinsey, 1988). In western countries the issues of 'social marketing' and corporate public relations also have major ethical components (Vinten, 1990). If companies seek to adopt Friedman-type principles, in which the profit motive and stakeholder satisfaction are high on the agenda, then there will inevitably be a knock-on effect for the marketing function.

Within such a framework, increasing market share 'at all costs' would be the order of the day and an accurate representation of product performance would be relegated in importance. Given the argument that

> We need a far better understanding of how individuals and, especially, organizations internalize codes of what is not acceptable practice

then it becomes important for us to

> focus on the organization, how organizational change is achieved and the relationship that individuals and organizations have with the surrounding society.
>
> *(Gray, 1990, p. 15)*

Within this context the marketing process, lying at the interface between the organization and its environment, is an important pillar of corporate responsibility (see Smith, Chapter 12 of this book).

What then lies at the heart of ethical marketing? Like many of the functional areas detailed in this book, marketing is dominated by issues of technical uncertainty and the role of expertise therein. The most basic tenet of ethical behaviour that should be required of business relates to a fairer representation of product performance. It should not be permissible for corporations to make spurious claims concerning product environmental performance. A second strand of the process would entail corporations restraining from making misleading representations about product performance. Within this context, the burden of proof should be on the corporation rather than the consumer and, whilst this process will become implicit within new product liability standards, it should be incumbent upon marketers to recognize the limits of such uncertainty. This would be particularly important within the public relations function and the marketing of service provision for environmental improvements. One can question the industry's ability and willingness to 'self-regulate' in this respect and perhaps the only effective solution would be to tighten the regulatory framework to force a paradigm shift towards greening on the sector. It is to a discussion of this issue that the final section now turns.

Marketing the Environment: Towards a Paradigm Shift

Whilst it can be argued that the emergence of green issues presents companies with significant marketing opportunities, the marketing and promotional strategies that they adopt will require careful planning and implementation if they are to derive maximum benefits. An organization that designs a better product will gain a competitive advantage and its usefulness and attractiveness will influence acceptability, but it is important to remember that consumers also buy product benefits and not just product features.

The marketing philosophy has always been based on the need to focus on customer needs and wants, and organizations have in the past generally designed products that will satisfy these demands. It has usually been sufficient to present a product to the market place that consumers perceived as providing benefits that enabled them to satisfy their basic needs or wants. Today's consumers, however, are no longer satisfied with this provision. Their purchasing behaviour has become more sophisticated and they are now demanding that the products or services that they buy not only fulfil specific functions but are not harmful to themselves or their environment.

Many manufacturers argue that they are responding to these demands and while some genuinely are, the majority are merely using advertising and promotional techniques to create a green image. This approach has succeeded in selling mass-produced goods to the public in the 1960s and 1970s but is now no longer as effective on the more sophisticated, environmentally conscious consumer operating in the 1990s. If producers of consumer goods are to remain competitive in the market place they will have to change from an 'image-based' approach to marketing to a more customer-focused one.

Within the current context of corporate responses to environmental concerns there is a growing need to move away from an *ad hoc* process of self-regulation towards a more structured and comprehensive series of environmental regulations for marketing and product labelling. Such a regulatory strategy needs to ensure that products are *effectively* labelled regarding their contents and that claims of improved environmental performance are independently verified. Similarly, claims concerning the ability to recycle the product's packaging must only be made if the infrastructure for such a programme is in existence at the point of sale. Whilst criticism of enforced legislation can always be made, corporate groups have expressed the view that such a move would be welcomed as it would provide the necessary standard to which the industry would need to work (Alvord, 1991). In addition, strong legislation would provide a powerful barrier to entry for new entrants and would reinforce the position of the stronger companies in the market; both elements should provide inducements to the current market leaders to lobby governments for such controls. The likely winners of a tighter regulatory framework will be the public and the environment, although the downside will undoubtedly be in terms of increased cost per unit product which will ultimately be passed on to the consumer.

Industry frequently cites the extra costs involved in changing to alternative technologies and raw materials as the main reason for their tardiness in actively responding to environmental issues. However, as long as perceived benefits are greater than perceived costs, consumers will buy the product. This has been amply demonstrated by the success of niche marketing in the 1980s. The problem for industry is not that it is not reacting to green consumer demands, but rather, it is responding to them by relying almost exclusively on promotional techniques. It has failed to realize that imaginative sales promotions and package designs are no longer enough to sell products. If they do not respond more proactively to green consumer demands, they will soon realize the power of the consumer to determine what is, and is not, sold in the market place. This is already the case in environmentally aware markets like Germany and Scandinavia and increasing in importance in North America and Japan.

Industry has a wider responsibility to the public interest because many of its operations have an impact on the environment and consequently on society. There is currently a powerful argument that suggests that this responsibility goes far beyond the profit motive and that industry must subordinate its actions to an ethical standard of conduct (see Sethi, 1975; Freeman and Gilbert, 1988). Within this process, the serious environmental claims that have been attributed to certain products during the last few years will need to be substantiated before being used within product promotion. What this requires is a fundamental shift in the dominant value paradigm within marketing and, more broadly, within business in general. Central to this process will be the role of marketing education.

Marketing education is characterized by a predominantly managerial approach and, whilst this is not too surprising, it has led to a process of asymmetry (Karlinsky, 1987). This asymmetry has resulted in a management-centred view of business wherein marketing theory has sought to allow managers to maximize business performance through an understanding of the behaviour of consumers. Karlinsky (1987) calls for a radical review of this process and advocates a redress of the current imbalance through symmetrization. The implication of this process is that the interests of consumers would be given equal standing with those of managers, with the latter being exposed to more radical ideas concerning the obligations of business within a societal context. However, symmetrization is not just a matter of providing otherwise ignorant people with information, thereby expecting them to make more advised judgements. Rather, it is about undertaking research from the perspective of consumers and providing them with the tools and techniques necessary for effective decision-making which have hitherto been reserved for managers. Whilst such an approach may seem to be a radical shift in perspective it would not be a unique approach within management. Procurement (often termed 'purchasing'), for example, has long sought to adopt the perspective of the consumer in developing management strategies. Within the context of our current discussions, a consumer-centred approach to marketing should lead to a more responsible and ethical approach to the problem.

The advent of a growing concern for environmental quality could be seen as symptomatic of an evolving culture. It is as if the wider environmental implications of consumption patterns are diffusing through the more economically advanced societies. In this context it is reasonable to believe that business schools will gain support in any moves they make in the direction of representing society's interests as well as those of the supplier in the process of marketing.

References

Alvord, J. (1991) The greening of business education in the UK. Paper presented at the 11th Annual International Conference of the Strategic Management Society, *The Greening of Strategy-sustaining Performance*, Toronto, October 23–6.

Anon (1990) Friendly to whom, *The Economist*, 7 April, p. 117.

Borden, A. R. (1991) *Elements of Marketing*, DP publications, London.

Flux, M. (1990) Industry and the environment, *The Business Economist*, Vol. 21, no. 2, Spring.

Freeman, R. E. and Gilbert, D. R. (1988) *Corporate Strategy and the Search for Ethics*, Prentice Hall, Englewood Cliffs, N. J.

Gray, R. H. (1990) Business ethics and organizational change, *Leadership and Organization Development Journal* Vol. 11, no. 3, pp. 12–21.

Gronroos, C. (1990) Marketing redefined, *Management Decision*, Vol. 28, no. 8.

Karlinsky, M. (1987) Changing asymmetry in marketing, in A. Firat, N. Dholakia and R. Bagozzi (eds.) (1987) *Philosophical and Radical Thought in Marketing*, Heath, Washington D.C.

Kinsey, J. (1988) *Marketing in Developing Countries*, Macmillan, London.

Kotler, P. (1990) *Principles of Marketing*, Prentice Hall, New York.

Maslow, A. (1954) *Motivation and Personality*, Harper and Row, New York pp. 80–106.

McMurdo, L. (1990) Listeria outbreak ... not many dead, *Marketing Week*, 13 April, p. 19.

Roddick, A. (1991) *Body and Soul*, Ebury Press, London.

Sethi, S. P. (1975) Dimensions of corporate social performance: an analytical framework, *California Management Review*, Vol. 17, no. 3, pp. 58–64.

Simons, T. (1991) The greening of business education in the UK, *see* Alvord.

Simmonds, I. G. (1990) No rush to go green, *Area*. Vol. 1, no. 22, pp. 384.

Smith, D. and Hart, D. (1991) The greening of business education in the UK, *see* Alvord.

Smith, C. and Sambrook, C. (1990) Dead end street, *Marketing*, 10 May.

Vinten, G. (1990) Business ethics: busybody or corporate conscience? *Leadership and Organization Development Journal* Vol. 11, no. 3, pp. 4–11.

8

THE GREENING OF RISK ASSESSMENT: Towards a Participatory Approach

Frank Fischer

Introduction

The problem of technological and environmental risk has become one of the major policy issues of our time. During the past two decades there has been a growing recognition that technological progress has brought with it many dangers. Indeed, there is now in various quarters of society a growing distrust of modern techno-industrial progress (National Research Council, 1989, pp. 54–71). In the face of nuclear power catastrophes, the death of oceans and rivers as a result of oil spills and other disasters, greater structural unemployment because of computerized work, pollution and the greenhouse effect, the pervasiveness of toxic wastes, the rise of cancer rates, and the like, we have come to recognize that one of the prices paid for this technological advance has been a dramatic increase in risk, or at least the awareness of risk (Slovic *et al.*, 1980; National Research Council, 1989). Such events have given both credibility and political influence to the environmental movement, which emerged in the 1970s as a response to these and other related events (Douglas and Wildavsky, 1982).

The number of people who believe that we live in the riskiest of times has increased dramatically over the past two decades. For these people, particularly those of the green persuasion, the world is on the brink of ecological disaster. Modern technology is seen to constantly generate new threats to the earth's life-support systems and thus in turn to the stability of social systems. Especially important to this argument are the synergistic effects of these problems. It is not just the appearance of new problems that causes concern, but rather the emergence of countless serious problems at the same time, e.g. over-consumption of the earth's energy resources, the ozone hole, the toxic waste problem, contamination of rivers and oceans, the dangers of nuclear radiation, the rise of cancer rates, the destruction of the rain forests, the rise of the earth's temperatures, escalating population growth and more. Even though people are seldom exposed to one risk in isolation from the others, there exists little empirical information on the interactive effects of these dangers. For such reasons, those who see a dramatic increase in risks call for tighter control over technology, including the abandonment of some technologies considered to be particularly risky (such as nuclear power and genetic engineering), and the need for the development and introduction of more environmentally benign technologies.

The result has been a much greater awareness of the impacts of modern technologies on the environment, coupled with a questioning of our blind acceptance of technological progress. Indeed, the more radical ecological groups in the environmental movement have raised basic questions about our very way of life. Proponents of the 'green' philosophy see the solution to lie in a return to smaller, less hierarchical technological systems, with a much greater role for people (Porritt, 1984). The radical nature of this environmental critique of the modern techno-industrial system and an outline of the movement's alternative conception of the good society is illustrated in Table 8.1.

The struggle over the question of technological and environmental risk has elevated the 'search for safety' to the top of the political agenda (Wildavsky, 1988). The quest for safety has emerged as one of the paramount political issues of our time, both as a prominent public concern and a leading topic in intellectual discourse (Fischer and Wagner, 1990). In Germany, for example, it has led sociologist Ulrich Beck (1986) to literally define the post-industrial society as the 'Risk Society'. Whereas in earlier periods people feared political turmoil resulting from imbalances in the distribution of goods, Beck argues that in contemporary society people are increasingly worried about the risks engendered by their own systems of production. Moreover, the growing awareness of risks among men and women also produces today a new and distinctive pattern of consciousness. The class awareness of the earlier period, Beck maintains, is increasingly displaced by the omnipresence of a risk awareness common to all. A nuclear accident, for example, recognizes no social boundaries; the same moribund consequences hit both the rich and the poor. No one, regardless of their wealth, can escape the risks of technological disaster. Ironically, as he explains it, these risks have in their consequences become socially just.

Acceptable Risk as the Techno-industrial Response

Such concerns have forced the proponents of large-scale technological progress, corporate and governmental leaders in particular, to pay much greater attention to the regulation of advanced technologies. Indeed, the future of many new technologies today is believed to virtually depend upon the ability of regulatory institutions to restore public confidence in them. Numerous industrial and political leaders, in fact, express worry about an emerging environmental philosophy that would oppose the introduction of all new technologies – even tested technologies – without *absolute* prior proof that they pose no risks, a view exemplified by such radical ecologists as Jeremy Rifkin (Tivnan, 1988).

Table 8.1

Techno-industrialism	*Green/ecology*
1. *Basic values*	
Aggressive individualism	A cooperative communitarian society
Pursuit of material goods	Emphasis on spiritual and non-material values
Rationality and technocratic knowledge	Intuition and understanding
Patriarchal values/ hierarchical structure	Post-patriarchal feminist values/decentralization
Unquestioning acceptance of technology	Discriminating development and use of technology
2. *The environment*	
Domination over nature	Harmony with nature
Environment managed as a resource	Resources regarded as strictly limited
High energy consumption/ nuclear power	Low consumption/renewable energy sources
3. *The economy*	
Economic growth and demand stimulation	Sustainability, quality of life, simplicity
Free market economy	Low production for local needs
High income differentials	Low income differentials
Production for exchange	Production for use
Capital-intensive production	Labour-intensive production
4. *Political organization*	
Centralization/economies of scale	Decentralization/human scale
Representative democracy	Participatory democracy

Adapted from Porritt (1984), pp. 216–17.

Frightened by the proliferation of toxic waste sites, the failure of nuclear power plants and catastrophic oil spills, to name just a few, the public has increasingly begun to simply say no. This opposition has been most explicitly expressed by the much discussed NIMBY (Not In My Back Yard) phenomenon, seen by business and governmental leaders as a fundamental threat to modern techno-industrial progress, in short, the western way of life.

The concern often reflects very concrete experiences, the nuclear energy industry being the example *par excellence* (Rosenbaum, 1985, pp. 227–34). What started out as one of the magnificent peacetime wonders of modern technology has been brought to a virtual standstill in the USA by the environmental movement and the assistance of such accidents as Three Mile Island and Chernobyl. Sweden, moreover, has considered a total phase-out of nuclear energy in the coming decade, thanks to powerful environmental agitation.

The primary response has been an attempt to shift this political discourse to the search for 'acceptable risk'. Supporters of the modern techno-industrial complex argue that risk must be seen as a mixed phenomenon, always producing both danger and opportunity. Too often, they argue, the debate revolves purely around potential dangers (all too often centring on high impact accidents with low probability, e.g. nuclear meltdowns or runaway genetic mutations). Risk-taking, in contrast, must be seen as necessary for successful technological change and economic growth, as well as the overall resilience and health of modern society (Wildavsky, 1988).

The basic strategy of industrial and scientific leaders has been to focus the risk debate on technical factors (Wynne, 1987; Schwarz and Thompson, 1990, pp. 103–20). The approach is grounded in the view that technological dangers have been grossly exaggerated (particularly by the Luddites in the environmental movement who are said to harbour a vested political interest in exploiting the public's fears). The result, it is argued, is a high degree of ignorance among the general public about technological risks. The layperson, they point out, tends to worry a great deal about the safety of air travel but thinks nothing of driving his or her car to the airport, which statistics demonstrate to be much more dangerous (Lopes, 1987).

To counter this uncontrolled expansion of 'irrational' beliefs the goal has been to supply the public with more objective (technical) information about the levels of risks themselves. That is, the 'irrationality' of contemporary political arguments must be countered with rationally demonstrable scientific data. The solution is to provide *more* information – standardized scientific information – to offset the irrationalities plaguing uninformed thinkers, i.e. the public. More recently, in fact, the task has become the focus of a new subspecialty of risk management known as 'risk communication' (National Research Council, 1989).

Improving Technical Objectivity: Scientific Risk Assessment

During the 1980s, scientifically based quantitative risk assessment was instituted by the Environmental Protection Agency as the mandated criterion for risk-related policy decision-making. Risk assessment was introduced to supply technical data for framing policy issues pertaining to environmental and health risks resulting from chemicals, pollutants, pesticides, toxic wastes, PCBs, etc. The methodology was designed to organize and manage a public issue that threatened to get out of control.

Risk assessment is a methodological strategy designed to supply a technically rational basis for centralized regulatory decision-making. As a scientific model of rational decision-making, risk assessment reflects an amalgam of managerial and engineering methodologies (Mazur, 1980). The goal of risk analysis is to provide objectively standardized quantitative information about the reliability of a technology's performance (Coppock, 1984, pp. 53–146). Towards this end, it defines risk in terms of the physical properties of a technology and its environment. Like the engineering model upon which it is largely based, risk assessment proceeds from a fundamental assumption: namely, that a technological system (defined as an integration of physical and human factors) can be rigorously defined with a unitary concept of technical rationality (Schwarz and Thompson, 1990, pp. 102–22).

This means, more specifically, that physical risks exist separately from the symbolic context in which they are socially situated, and that social perceptions of technological systems and their risks can be strictly excluded. Indeed, social perceptions are 'irrationalities' that have little or nothing to do with technical knowledge. The goal, in short, is to isolate and measure the objective probabilities of technical failure in terms of intrinsic and extrinsic physical properties.

In analytical terms, risk is expressed as the product of the estimated degree of harm (death or damage) a given technical failure would cause, and its probabilities of occurrence (Wynne, 1987; Irwin, Smith and Griffiths, 1982). In complicated technological systems, this involves first breaking the system down into various measurable components – materials, pipes, pumps, seals, coolants, fail-safe mechanisms, etc. – and then, second, measuring the statistical probability of a failure of each of these dimensions, using past performance data, experimental assessments and expert judgements. Third, it means an examination of crucial environmental factors that might either precipitate or exacerbate a technical failure, such as geological fault lines (which might, for example, cause earthquakes) or climatic conditions such as wind patterns (which would influence the spread of dangerous particles). Fourth, the foregoing factors, stated statistically as multiple probabilities, must be integrated through a modelling process based on decision-oriented event and fault trees. After calculating the various chains of probabilities to provide an overall estimate of system failure, the figure must be multiplied by the estimated damages. For example, the health effects of human exposures to the release of dangerous substances, such as chemicals or radioactive particles, must be estimated using epidemiological, medical and actuarial data to determine the long- and short-run damages to human well-being, latent as well as delayed.

The result is a set of objective statements that are to be the focus of regulatory deliberations about risk, thus replacing risk 'perceptions' divined, so to speak, by technically uninformed social actors, particularly those who lead environmental movements (Andrews, 1990, pp. 167–86).

Methodological Critique: the Social Dimension

The method has proved to be anything but a success. Although it has produced a mountain of quantitative data, it has altogether failed to reassure the public. Indeed, in many ways it has only worsened the situation. The fundamental problem is to be found in the technical framing of the risk problem.

Given the complexity of technological systems, the task of generating quantitatively standardized estimates of risk requires the use of very narrowly defined empirical assumptions and analytical concepts, including the very definition of what constitutes a technology. This is not the place to go into technical details (concerning such matters as data limitations, units of analysis, the security of safety components, levels of professional expertise, and the like). Suffice it to say that analysts are forced to make many uncertain assumptions – sometimes even heroic assumptions – which are later masked by the precision of their risk statements. As Wynne (1987) puts it, the price paid for precision is often a high degree of hidden ignorance about the safety of technical systems.

This glossing over of empirical and analytical uncertainties has given rise to a good deal of disagreement among the experts themselves. Indeed, their judgements about these statistics often range from completely reliable to totally useless. Needless to say, this display of technical disagreements – featuring counter-experts criticizing experts – does nothing to assuage the public of its fears (Collingridge and Reeve, 1986). Indeed, there is little doubt that the open display of disagreement among experts has only heightened public worries. To put it bluntly, risk asssessment – on its own terms – has not only failed as a technical tool, it has also in the process exacerbated the very doubts it set out to assuage (Schwarz and Thompson, 1990; Wynne, 1987).

This reluctance to directly consider contextual factors, however, breeds major problems. The method is greeted with public distrust, even suspicion. It comes to be seen – correctly – as a managerial decision methodology strategically employed to gloss over and hide important social and political issues. The result is a political legitimacy problem. By denying legitimacy to the values and anxieties that arise from the social contexts in which technologies are situated, those who seek to regulate risk seriously jeopardize their own credibility by saying to people that their social experiences and searches for meaning do not count (Wynne, 1987; Schwarz and Thompson, 1990).

The Large-scale Technological System as Social Phenomenon

In contrast to the engineering orientation, then, the regulatory task of risk assessment in a democratic society has to approach its object of investigation in a very different way, which brings the discussion to the second point. The first step towards a solution is to be found in the theoretical redefinition of large-scale technological systems. Rather than a complicated set of nuts and bolts, we must recognize such systems to be a special type of social phenomenon (Joerges, 1988). In actuality, large-scale technological systems are integrated sets of techno-institutional relationships embedded in both historical and contemporary social processes. They are complicated technical processes functionally woven together by networks of socio-organizational controls (Perrow, 1984).

This reality, in fact, bears directly on both the empirical estimation of risk and the social perceptions of acceptable risk. With regard to the social perception of risk, much to the chagrin of risk analysts, these institutional factors are discovered to play an essential role in the layperson's perceptions of technological systems. Social perceptions and understandings of large technological systems – those of workers and citizens in particular – are fundamentally rooted in their concrete social experiences with decision-making institutions and their historically conditioned relationships.

Any single technical 'event' or 'decision' is in fact located within a continual socio-institutional process, which is itself an integral dimension of the large-scale technological system. For example, sociological evidence (as well as common sense) shows that workers cannot altogether divorce their responses to physical risks from their attitudes towards social relations in the plant, particularly those pertaining to managerial practices. If the workers' social relations with management are pervaded by mistrust and hostility, the ever-present uncertainties of physical risks in the plant are amplified (Wynne, 1987).

For risk analysts, this amplification is a subjectively irrational behaviour (sometimes portrayed as rooted in a neurotic pathology). The cynics among them see this inclusion of social perceptions to result from a simple fact: namely, that people rely on this social dimension as it is the only aspect of such systems that they are capable of understanding (Wynne, 1987; Douglas and Wildavsky, 1982). This, in fact, has led to a line of behavioural research concerned with the psychology of risk perceptions (Slovic, 1984). Researchers seek to determine how and why people attach social meanings – supposed irrational meanings – to specific technologies. Why, for example, has nuclear power – in the face of favourable risk assessments – attracted such negative images? Why do they express concern about chemical installations when the levels of risk as determined by technocrats are quite low?

But really how 'irrational' is this behaviour? Which brings us to the third point. Here the most obvious empirical perception pertinent to this issue needs to be underscored: namely, the fact that over the past fifteen years we have more and more recognized the sources of technological hazards and catastrophes to be frequently the result of institutional failures. For example, the tragedy at the Bhopal chemical plant was discovered to be the fault of workers who ignored the misalignment of valves (Shrivastava, 1992); the Chernobyl catastrophe was caused by plant operators who overlooked established procedures (Medvedev, 1991), while their counterparts at Three Mile Island applied them inappropriately (Perrow, 1984); the Prince William Sound oil spill in Alaska resulted from neglect of standard navigation practices by the *Exxon* tanker captain (Keeble, 1991); the failure of the space shuttle *Challenger* occurred as a consequence of management's unwillingness to heed the warnings of company engineers (Lambright, 1989; Vaughan, 1990); a serious plane crash at Los Angeles International Airport was caused by air traffic controllers who cleared two planes to use a runway at the same time. The examination of such accidents shows them to be the consequences of untrustworthy or irresponsible organizational and managerial systems. Indeed, it becomes increasingly difficult to find crises in which this is not at least partly the case (Clarke, 1989; Dobrzynski, 1988; Paté-Cornell, 1990). Such events have scarcely contributed to the public's trust and confidence in management's ability to reliably ensure personal and environmental protection.

Institutional and managerial factors are thus themselves very real sources of technological risk and uncertainty and people often correctly perceive this, even if formal risk assessment chooses not to. As Wynne (1980, p. 188) nicely puts it, 'uncertainties which are not acknowledged and dealt with full-frontally have a way of creeping back into social perceptions, perhaps dressed in a different language.' From this perspective, social perceptions can just as easily be seen as sources of experiential knowledge that must be taken seriously. This is particularly the case when they pertain to the kinds of situational circumstances that escape the broad social generalities typically sought by risk perception research.

Rather than being a technical issue plagued by social perceptions, the risk problem turns out to be as much a social question related to technical issues. Not only does the abstract language of risk assessment underplay the institutional-managerial context of technological systems, it also fails to recognize that they reflect larger normative concerns rooted in the society itself (Schwarz and Thompson, 1990). Some, in fact, argue that a lack of credibility in our institutions and leaders is, in reality, the central factor driving the fear of technological systems, an argument central to the environmental movement.

Towards a Participatory Approach

One should not underestimate the problem posed here. The search for a more democratic form of risk assessment confronts a number of the most sophisticated political and epistemological problems of our time. In epistemological terms, it is nothing less than the question of how to relate empirical data to norms and values – a very old question that continues to occupy philosophers (Hawkesworth, 1988). In political terms, it asks how we are to transform our increasingly techno-bureaucratic institutions into less hierarchical democratic structures (Fischer, 1990). Indeed, in an age of technocratic expertise the question is critical: one can justifiably argue that the very future of democracy depends on it (Petersen, 1984).

Within the scope of this chapter I cannot, of course, adequately supply answers to these challenging questions. I will thus only attempt to point risk assessment in a direction suggested by the foregoing critique. The first step, which should be apparent by now, is to move beyond the idea that technical knowledge, coupled with improved communications of empirical findings, can alone answer questions posed in the social and political world. Put more bluntly, it is time to finally recognize this to be the technocratic ideology that it is (Fischer, 1990)!

The challenge is to find ways to more comprehensively and meaningfully integrate technical and social data, in both analytical inquiry and public discussion. In the case of the latter, I mean something more than calling for new forums in which experts present and discuss their findings with the public, largely the contemporary approach to risk communication (National Research Council, 1989). In this scenario, science presents something of a *fait accompli*. By virtue of their 'exalted' method, scientists command the privileged position in such forums; the intimidated public is relegated to the role of passive listener largely closed out of the discussion (Smith, 1990). In the end, as much experience shows, the process only further breeds the kind of alienation that already presents risk assessment with its legitimacy problem (Nelkin, 1984, pp. 18–39; Petersen, 1984, pp. 1–17).

The solution is to be found in the invention of both new methodological approaches and institutional forums capable of opening up the risk assessment process to non-experts. Laypersons must be integrated into the process as part of a discussion of the social and institutional issues upon which quantitative/technical risk assessment calculations rest. This means building the much needed social discourse *into* the phases of scientific research itself (Eldon, 1981). Consider some more specific illustrations.

At the outset of risk assessment, a wider range of stakeholders must be built into the initial discussions of what the risk problem is in the first place; they must again be built into the search for risks; then again when it comes to the determination of the relative importance of risks and benefits and how they might be most meaningfully quantified and measured; and finally their perspectives on the interpretations of the resulting risk estimates must be elicited. It is important to elaborate somewhat on each of these phases.

First, building the participants into decisions about problem definition is essential to an effective strategy of risk assessment, especially once we recognize the fundamentally social nature of technological systems. Important in the definitional problem, for instance, are questions concerning what the technology actually is, what are its significant components and connections, and what are its boundaries and external contexts.

For instance, in a complex techno-institutional system such as hazardous waste disposal (involving a cycle of activities ranging from production, collection and transportation to treatment and storage), how do we determine which actors and institutions are in the system and which are out? In some cases, of course, the answer is easy: the incinerator at the point of disposal, for example, is an essential component of the system. But what about the production process? How much of that, if any, is to be included in the definition of the hazardous waste system? Some argue that we cannot adequately understand the process without including the industry that produces the waste, although analysis typically ignores this dimension.

Perhaps even more critical, once we recognize that the social dimension is itself a fundamental source of risk, it is again important to bring the social actors themselves into the process of identification and search for risks. In particular, workers familiar with the everyday operations of a plant have important experiential knowledge about how the socio-institutional structures and processes of the system actually function. Such concrete knowledge is of critical importance in the search for alternative sources of risk. Information of this type often cannot be obtained from management and it certainly cannot be gleaned from the kinds of abstract statistical analysis that has typically characterized risk assessment (Wynne, 1987).

While the assignment of weights to risks and benefits sounds like a highly technical task, this process too is grounded in a large number of normative considerations. How, for example, are risks and benefits to be counted? If a benefit is intangible or not traded in markets, how should we establish price values for it? If the benefits occur far into the future, how should their values be 'discounted' to accord with the fact that money available only in the future is typically worth less than money available immediately?

Or, how does one attach numbers to the value of lives saved by a safety procedure? Should the value of each life be the expected future earnings of each person? Should we count as part of the programme's costs all the future medical expenses of the people saved? Should we add in their children's schooling and medical costs? Should we count as benefits the taxes paid by the people we rescued?

Finally, citizens and workers must have important inputs into the process of interpreting the meaning and uses of an analysis. Especially important, in this respect, is the relationship of the findings to the specific situational circumstances to which they are to be applied. The production of generalized information about the risks of a technological system can never be more than guidelines that must be interpreted within the specific contexts to which they are applied. Wynne (1987), for example, has shown how centrally established regulatory rules about risk often produce enormous problems in their implementation. Because each technological system is located in a different geographical area with varying physical and social characteristics, there is no possibility for the rule to apply directly to specific facilities. In each case, then, the meaning of the data must be socially negotiated between the central authorities and the local officials. Without such normative negotiations, central rules are seen to often do more harm than good. Not only do they create a great deal of frustration among local participants, they also themselves become the sources of risks.

These are typical of the kinds of normative questions that underlie an otherwise technical analysis. The processes of empirical measurement, data collection and mathematical analysis – the processes typically identified as scientific risk assessment – are seen to be founded on a very wide range of crucial normative judgements. Rather than questions and issues merely brought to the scientific process by concerned social and political participants, they are in fact an inherent part of scientific risk assessment. Although scientists must grapple with these questions, such issues can only be established by social judgements. In this realm, scientists have no epistemologically privileged position over the other members of society, although in fact they frequently make such judgements in the name of science (Stone, 1988, pp. 194–200). These are societal questions and can only be legitimately dealt with through societal processes.

The advantage of such an integration rests on two fundamentally important contributions. First, it builds into the analytical process the stakeholders' pragmatic experiential knowledge about technical and institutional risks; and, second, it addresses the essential issues of public legitimation and motivation. By involving them in these normative dimensions of the scientific process as it proceeds, stakeholders become cooperative participants in the formation of scientific arguments, rather than mere passive listeners, the result of which is greater commitment to the analytical conclusions, i.e. the very problem risk assessment has been unable to resolve.

Participatory Expertise

For those in the technical and managerial sciences who take democracy seriously this constitutes an important challenge. First, and perhaps foremost, a commitment to the furtherance of democratic participation and public empowerment must be a matter of professional commitment. Such a commitment, of course, is not the sort of thing most scientists are accustomed to, or even willingly welcome, as it clearly represents a challenge to their privileges and status. But, as the sociology of science has made clear, social commitments are already imported into the research process. A commitment to participation is therefore as much a matter of changing a social practice as it is of disturbing the epistemological requirements of science (Weingart, 1990).

But equally important is the need for institutional and methodological innovations. Risk assessment, as a managerial decision science, is a product of a bureaucratic system; it is a tool designed to guide hierarchically structured decision-making processes. A more democratically structured practice of expertise would similarly require a more participatory set of institutions. This means new political innovations that can only be brought about through political struggles, particularly of the type advanced by participatory-oriented social movements, such as the ecology movement, grounded in a very different set of values (Friedmann, 1987; Fischer, 1990).

Experimentation with such participatory structures for risk assessment is not an altogether new idea in the USA. In the 1970s there were a wide range of experiments in this direction. Important in this respect was the rise of the public interest science movement and such institutions as the Center for Science in the Public Interest.[1] To support such efforts the National Science Foundations established a programme to foster the development of citizen-oriented science (Hollander, 1984). Largely judged to be successful, the programme was designed to help develop and facilitate efforts that sought to incorporate diverse points of view into science and technology projects, thus helping to assure the representation of all groups with substantial interests in such matters.

Regrettably, such innovative efforts were beaten back by scientific lobbies and conservative politicians during the Reagan years of the 1980s. The argument was always the same: the layman simply cannot understand and responsibly judge complex technological issues. Such participation is said to be unrealistic, if not utopian. As a former president of the National Academy of Sciences put it:

> most members of the public usually don't know enough about any given complicated technical matter to make meaningful judgements. And that includes scientists and engineers who work in unrelated areas.
>
> *(Handler, 1980)*

There is truth in such statements, but they tend unfortunately to refer to the purely technical aspects of such problems, obscuring the wide range of normative social judgements upon which such technical matters depend. Even more important, numerous experiments show this judgement to be premature (Peterson, 1984).

Another step in this direction worth consideration would be the establishment of publicly financed centres for the study of technological impacts, located perhaps within the public university systems (Plotkin, 1989). These could serve as training grounds and laboratories for democratically committed engineers, scientist and risk specialists. They would be centres for the development of an independent and democratically informed version of science that could be used to review and analyse major techno-industrial developments and the risks they pose. They could provide the informed technical capacity to make a more democratic case for opposing, modifying or retaining complex technological risks on the basis of public interests.

In methodological terms, what is needed is an approach capable of facilitating the kinds of discussion that go on in such participatory contexts. The method would provide a format and a set of procedures for organizing the interactions between experts and the laypersons they would seek to assist. One promising movement in this direction is the development of 'participatory research' (Fischer, 1990). A methodology designed primarily for research problems characterized by a mix of technical and social factors, participatory research is a practice that has begun to take shape among alternative social movements, including alternative technology and ecological movements, especially in the Third World (Society for Participatory Research in Asia, 1982).

To many conventionally trained scientists, both physical and social, the idea of participatory research often sounds outrageously unscientific, but, in actuality, it is in most ways only the scientific method made more time-consuming and perhaps more expensive, at least in the short run. Fundamentally, it is a more progressive version of something already accepted as a methodology in the managerial sciences, namely action research (Argyris, 1985). Like action research, participatory research is designed as a methodology for integrating social learning and goal-oriented decision-making. Where the former was eventually co-opted by the managerial sciences to serve the rather narrowly defined needs of bureaucratic reform (typically defined as participatory management), participatory research is largely an effort to carry through on action research's earlier commitment to democratic participation (Reason and Rowan, 1981; Fernandes and Tandon, 1981; Kassam and Mustafa, 1982; Merrifield, 1989).

In the context of socio-technical problems, such as those confronted in risk assessment, participatory research has been put forward as an effort to gear expert practices to the requirements of democratic empowerment. Rather than providing technical answers designed to bring political discussions to an end, the task is to assist citizens in their efforts to examine their *own* interests and to make their *own* decisions (Hirschhorn, 1979). Towards this end, it conceptualizes the expert as a 'facilitator' of public learning and empowerment. Beyond merely providing data, the facilitator must also become an expert in how people learn, clarify and decide for themselves (Fischer, 1990). This includes coming to grips with the basic languages of public normative argumentation, as well as knowledge about the kinds of environmental and intellectual conditions within which citizens can formulate their own ideas. It involves the creation of institutional and intellectual conditions that help people pose questions and examine technical analyses in their own ordinary (or everyday) languages and decide which issues are important to them.

Participatory research, it should be emphasized, is not appropriate for all scientific problems. There is no need for physicists studying the structure of the atom to consult the layman. For those to which it does apply, it encounters many difficulties (Eldon, 1981). Most of the problems, however, are social and interpersonal rather than scientific *per se*: It takes, in short, a lot of time, political commitment and interpersonal skill to build people into complex decision processes.

But its practitioners also point to two important pay-offs. First, it identifies very real and important dangers that hide behind the generalities buried in the technocrats' calculations. That is, it brings to the fore the very problems that have been overlooked by technocratic risk assessment. Second, and every bit as important, participation in decision-making helps to build both credibility and acceptance of research findings (Dutton, 1984; Friedmann, 1987), the most critical failure facing the contemporary risk assessment approach.

Concluding Remarks: Possibilities and Prospects

There can be no doubt that such institutional and methodological proposals will be vigorously resisted by many powerful industrial and political interests. Industrial leaders, politicians and scientists will criticize it as being too expensive, time-consuming, inefficient and unscientific (the first three of which are surely true in the short run). Others will lambast the idea of participatory expertise as being fundamentally out of sync with the organizational practices of contemporary corporate capitalism, which indeed it is (certainly large-scale corporate-bureaucratic capitalism).

In this respect, the kind of participation proposed here is much more than a matter of managerial reform. First and foremost, it is a call for radical democratic changes in organizations more generally. Running counter to the workings of the political and market structures that constitute contemporary western capitalist systems, particularly the practices of private ownership and hierarchical managerial control, participatory expertise is ultimately part of an alternative way of life. As was made clear in the discussion of the green challenge, a serious commitment to participatory expertise can, in fact, only have genuine meaning when systematically incorporated into the operational practices of an alternative organizational life.

Against the scope of these requirements, the prospects of participatory expertise will strike many as being quite bleak. The situation, however, is not as hopelessly entrenched as it at first appears. Perhaps most important, we can count on business leaders to follow their own economic self-interest. When economically threatened by environmental challenges, businesses most certainly take notice. The growing severity of the environmental threat, coupled with a greater environmental consciousness on the part of both stockholders and the public, has already led many businesses to confront the market consequences of neglecting appropriate action (Epstein, 1991, p. 11; Holusha, 1991, p. 1). As businesspeople more and more recognize that the technocratic approach to risk is itself responsible for exacerbating – if not creating – many of their most serious public conflicts over environmental matters, they will increasingly reassess their commitment to technocratic management. In fact, once they see participation to be the key to the public legitimation of technologies, and thus a mechanism for avoiding costly environmental débâcles (such as the abandonment of nuclear power plants after spending billions to construct them), experimentation with participatory approaches will actually begin to look attractive.

Beyond these critical steps of awareness, the problem is in significant part a matter of experimenting with many of the approaches already available. Participatory expertise, as indicated earlier, is largely only a radicalization of a practice long included in the managerial tool-box, namely action research. Action research, along with the concept of participatory management (quite popular in progressive managerial circles in the 1970s), has laid sufficient intellectual groundwork for coming to grips with participatory expertise. The task is essentially to extend the application of the methodology beyond the restricted confines of managerial decision-making to include both a broader range of employees and the relevant public groups. With regard to employees, there exists a substantial amount of untapped information about experiments in organizational democracy, particularly those based on the experiments of self-managed enterprises, that provides very useful insights about worker participation and, to some extent, the role of the professional experts in such settings.

With regard to bringing in greater community involvement, there are also a growing number of experiments that point to interesting possibilities. Perhaps the most significant from the perspective of a participatory approach to risk assessment is a California experiment with 'siting contracts'. The siting contract approach is largely designed to confront the very sort of political clash that now regularly accompanies hazardous projects. Essentially it is an effort to bring together technocratic risk assessors (who tend to focus on scientifically predicting potential problems with plant equipment and operations in advance of their utilization) and community groups (who worry more about uncovering dangerous conditions that might emerge later during the operation of the plant). Through a formalized set of power- and risk-sharing procedures, management agrees to open its doors to the host community for review of operating and safety procedures, while, in turn, the community agrees to formally accept and sanction the facility, if satisfied that it is safe (Elliott, 1988). Short of acceptance, the contract provides for a discontinuation of operations.

By opening the facility for review, as well as establishing a regularized discussion relationship between plant managers and community leaders, the siting contract thus sets in motion a process of detection and mitigation of the kinds of dangers that block public acceptance of such projects. As Mazmanian and Morell (1990, p. 138) put it, 'inviting the community to participate in the oversight of a facility is obviously a radical departure from past practices, yet it is clearly an inclusive, community-based and democratic process.'

If managers must recognize the need to take seriously the participatory requirements of the ecological challenge, the ecological movement must also recognize and accept the need to confront more directly the tasks of management. Beyond protest and critique, the future of the movement will ultimately depend on its ability to translate its theoretical prescriptions into organizational and professional-managerial practices. As Hajer (1993, p. 35) has shown in the case of acid rain in Great Britain, environmental groups can win the public debate but fail to change policy. Because technocrats have succeeded in institutionalizing their principles in the specific organizational procedures and expert assessment techniques that guide Britain's environmental ministries, they have managed to maintain their control of environmental policy even after having been defeated by environmentalists in public forums.

In so far as the 'real-world' clash between the ecological and technocratic paradigms must as well play itself out at this operational level, environmental theorists cannot afford to ignore these more practical issues of policy analysis and management. To adequately counter the technocratic position they must specify more clearly the ways in which democratic participation can be made compatible with the technical realities of such practical tasks. They must, in short, begin to concentrate on the 'nuts and bolts' of participatory practices (Fischer, 1991, p. 35).

An important step in this direction must be the establishment of alliances with progressively oriented business leaders willing themselves to experiment with participatory solutions and strategies. The goal must be the development of a cadre of green managers and risk professionals capable of developing an ecologically sound approach to business and management. For environmentalists the point is crucial. The future of the cause will ultimately depend on the movement's ability to translate broad political perspectives into concrete organizational and analytical practices. Nothing, in this respect, can have a higher priority than the greening of risk assessment.

Acknowledgement

This paper expands on and modifies the argument presented in the author's paper 'Risk assessment and environmental crisis', *Industrial Crisis Quarterly*, Vol. 5, no. 2. pp 119–24. © 1991 Elsevier Science Publishers BV.

Notes

1. One of the most noteworthy examples in the USA has doubtless been the Cambridge Experimental Review Board (Knox, 1977). In this case the city of Cambridge, Massachusetts appointed a broadly representative non-scientific commission to assess

the risks involved in allowing Harvard University and Massachusetts Institute of Technology, both within the city's boundaries, to experiment with Recombinant DNA. After many months of intensive work, the group issued a report that stunned the scientific community. As one Harvard Nobel Prize-winner and critic of the commission put it, the document was 'a very thoughtful, sober, conscientious report' (Dutton, 1984, p. 148). It was widely agreed, regardless of what one thought of the commission's final verdict (which largely supported the continuation of the research), that a representative group of non-scientists could grapple with a profoundly complex science policy issue and develop recommendations widely seen to be intelligent and responsible.

References

Andrews, Richard N. L. (1990) Risk assessment: regulation and beyond, in Norman J. Vig and Michael E. Kraft (eds.) *Environmental Policy in the 1990s*, Congressional Quarterly Press, Washington, D.C., pp. 211–34.

Argyris, Chris (1985) *Action Science*, Harvard University Press, Cambridge, Mass.

Beck, Ulrich (1986) *Risiko Gesellschaft*, Surkamp, Frankfurt/Main.

Clarke, Lee (1989) *Acceptable Risk: Making Decisions in a Toxic Environment*, University of California Press, Berkeley.

Collingridge, D. and Reeve, C. (1986) *Science Speaks to Power*, St Martins Press, New York.

Coppock, R. (1984) *Social Constraints on Technological Progress*, Gower, Hampshire, UK.

Dobrzynski, J. (1988) Morton Thiokol: reflections on the shuttle disaster, *Business Week*, 14 March, pp. 82–91.

Douglas, M. and Wildavsky, Aaron (1982) *Risk and Culture*, University of California Press, Berkeley, Calif.

Dutton, Diana (1984) The impact of public participation in biomedical policy: evidence from four case studies, in James C. Peterson (ed.), *Citizen Participation in Science Policy*, University of Massachusetts Press, Amherst.

Eldon, Max (1981) Sharing the research work: participatory research and its role demands, in Peter Reason and John Rowan (eds.) *Human Inquiry: A Sourcebook of New Paradigm Research*, Wiley, New York, pp. 253–66.

Elliott, Michael P. (1988) The effect of differing assessments of risk in hazardous waste siting negotiations, in Gail Bingham and Timothy Mealey (eds.) *Negotiating Hazardous Waste Facility Siting and Permitting Agreements*, Conservation Foundation, Washington, D.C.

Epstein, Marc J. (1991) What shareholders really want, *New York Times*, 28 April, Section 3.

Fernandes, Walter and Tandon, Rajesh, (1981) (eds.) *Participatory Research and Evaluation: Experiments in Research as a Process of Liberation*, Indian Social Institute, New Delhi.

Fischer, Frank (1990) *Technocracy and the Politics of Expertise*, Sage, Newbury Park, CA.

Fischer, Frank and Wagner, Peter (1990) (eds.) Technological risk and political conflict: perspectives for West Germany, *Industrial Crisis Quarterly*, Vol. 3, pp. 149–54.

Fischer, Frank (1991) Risk assessment and environmental crisis, *Industrial Crisis Quarterly*, Vol. 5, pp. 113–32.

Fisher, Walter R. (1984) Narration as a human communication paradigm: the case of public moral argument, *Communications Monographs*, Vol. 51, March, pp. 1–22.

Friedmann, John (1987) *Planning in the Public Domain*, Princeton University Press.

Hajer, Maarten (1993) Discourse-coalitions and the institutionalization of policy practices: the case of acid rain in Britain, in Frank Fischer and John Forester (eds.) *The Argumentative Turn in Planning and Policy Analysis*, Duke University Press, Durham, NC.

Handler, P. (1980) In science, 'no advances without risks', *US News and World Reports*, 15 September, p. 60.

Hawkesworth, M. E. (1988) *Theoretical Issues in Policy Analysis*, State University of New York Press, Albany, NY.

Hirschhorn, Larry (1979) Alternative service and the crisis of the professions, in John Case and Rosemary C. R. Taylor (eds.) *Coops, Communes and Collectives: Experiments in Social Change in the 1960s and 1970s*, Pantheon, New York.

Hollander, Rachelle (1984) Institutionalizing public service science: its perils and promise, in James C. Peterssen (ed.) *Citizen Participation in Science Policy*, University of Massachusetts, Amherst, pp. 75–95.

Holusha, John (1991) Hutchinson no longer holds its nose: at 3M, cleaning up pollution has become the corporate ethic, *New York Times*, 3 February, Section 3.

Irwin, Alan, Smith, Denis and Griffiths, Richard (1982) Risk analysis and public policy for major hazards, *Physics and Technology*, Vol. 13, pp. 258–65.

Joerges, Bernward (1988) Large-scale technical systems: concepts and systems, in R. Mayntz and T. P. Hughes (eds.) *The Development of Large Technical Systems*, Westview Press, CO, pp. 9–36.

Kassam, Yusuf and Mustafa, Kemal (1982) (eds.) *Participatory Research: An Emerging Alternative in Social Science Research*, African Adult Education Association, Nairobi.

Keeble, John (1991) *Out of the Channel: the Exxon Valdez Oil Spill in Prince William Sound*, Harper Collins, New York.

Knox, Richard (1977) Layman is center of scientific controversy, *Boston Globe*, 9 January.

Lambright, H. W. (1989) Governmental-industry relations in the context of disaster: lessons from *Apollo* and *Challenger*. Paper presented at the Second International Conference on Organizational Crisis Management, New York University, Stern School of Business.

Lopes, L. L. (1987) The rhetoric of irrationality. Paper presented at the Colloquium on Mass Communications, University of Wisconsin, 19 November.

Mazmanian, Daniel and Morell, David (1990) The NIMBY syndrome: facility siting and the failure of democratic discourse, in Norman Vig and Michael Kraft (eds.) *Environmental Policy in the 1990s*, Congressional Quarterly Press, Washington, D.C.

Mazur, Allan (1980) Social and scientific causes of the historical development of risk assessment, in Jobst Conrad (ed.) *Society, Technology and Risk Assessment*, Academic Press, New York, pp. 151–7.

Medvedev, Grigori (1991) *The Truth About Chernobyl*, Basic Books, New York.

Merrifield, Juliet (1989) *Putting the Scientists in Their Place: Participatory Research in Environmental and Occupational Health*, Highlander Center, New Market, TN.

Mink, L. P. (1978) Narrative form as cognitive instrument, in R. H. Canary and H. Kozick (eds.) *The Writing of History: Literary Form and Historical Understanding*, University of Wisconsin Press, Madison.

National Research Council (1989) *Improving Risk Communication*, National Academy Press, Washington, D.C.

Nelkin, Dorothy (1984) Science and technology policy and the democratic process, in James C. Petersen (ed.) *Citizen Participation in Science Policy*, University of Massachusetts Press, Amherst.

Paté-Cornell, E. (1990) Organizational aspects of engineering system safety: the case of offshore platforms, *Science*, 30 November, pp. 1210–17.

Perrow, C. (1984) Normal accidents, in Frank Fischer and Carmen Sirianni (eds.) *Critical Studies in Organization and Bureaucracy*, Temple University Press, Philadelphia, PA, pp. 287–305.

Petersen, James C. (1984) Citizen participation in science policy, in James C. Petersen (ed.) *Citizen Participation in Science Policy*, University of Massachusetts Press, Amherst.

Plotkin, Sidney (1989) RATs to technology! Conflict, power and the crisis of industrial siting, *Industrial Crisis Quarterly*, Vol. 3, pp. 1–16.

Porritt, Jonathon (1984) *Seeing Green*, Basil Blackwell, Oxford.

Reason, Peter and Rowan, John (1981) *Human Inquiry: A Sourcebook of New Paradigm Research*, John Wiley, New York.
Rosenbaum, W. A. (1985) *Environmental Politics and Policy*, Congressional Quarterly Press, Washington, D.C.
Shrivastava, Paul (1992) *Bhopal: Anatomy of a Crisis*, 2nd ed. Paul Chapman, London.
Schwarz, M., Thompson M. (1990) *Divided We Stand*, Harvester and Wheatsheaf, Hertfordshire.
Slovic, Paul (1984) Behavioral decision theory: perspectives on risk and safety, *Act Psychologica*, Vol. 56, pp. 183–203.
Slovic, P., Fischoff, B. *et al.* (1980) Facts and fears: understanding perceived risk, in R. Schwing and W. A. Albers (eds.) *Social Risk Assessment: How Safe is Safe Enough?* Plenum, New York, pp. 75–98.
Smith, Denis (1990) Corporate power and the politics of uncertainty: conflicts surrounding major hazard plants at Canvey Island, *Industrial Crisis Quarterly*, vol. 4, no. 1, pp. 1–26.
Society for Participatory Research in Asia (1982) *Participatory Research: An Introduction*, Society for Participatory Research in Asia, Rajkamal Electric Press, 4163 Arya Pura, Delhi 110007.
Stone, Deborah A. (1989) Causal stories and the formation of policy agendas, *Political Science Quarterly*, Vol. 104, no. 2, pp. 281–300.
Stone, Deborah A. (1988) *Policy Paradox and Political Reason*, Scott Foresman, Glenview, Ill.
Tivnan, E. (1988) Jeremy Rifkin just says no, *New York Times*, 18 October p. 38.
Vaughan, D. (1990) Autonomy, independence and social control: NASA and the space shuttle *Challenger, Administrative Science Quarterly*, Vol. 35, no. 3, pp. 225–57.
Weingart, Peter (1990) Science abused – challenging a legend. Paper presented at the Seminar on Sociology of Science at the Inter-University Centre, Dubrovnik, 7–15 May.
Weiss, Andrew (1990) Scientific information, causal stories and the ozone hole controversy. Paper presented at the Seminar on the Sociology of Science, Inter-University Centre, Dubrovnik, 7–15 May.
Wildavsky, A. (1988) *Searching for Safety*, Rutgers University Press, New Brunswick, N. J.
Wynne, Brian (1980) Technological risk and participation under uncertainty, in J. Conrad (ed.), *Society, Technology and Risk Assessment*, Academic Press, New York.
Wynne, Brian (1987) *Risk Management and Hazardous Waste: Implementation and the Dialectics of Credibility*, Springer, Berlin.

9

MANAGING ENVIRONMENTAL IMPROVEMENT WITHIN A MAJOR CHEMICAL COMPLEX

Geoff Essery

Introduction

We all want to live in a green and pleasant land and most of us also want the benefits that industry brings to the quality of life. We want to go on eating a varied diet. We want to carry on wearing comfortable easy care clothes and having detergents to wash them time after time. We want to continue living in homes or going to offices, schools and shops that are warm, well decorated and well maintained. We appreciate the convenience of hopping into the car or on to public transport to get about. We expect that when we are sick, there will be pharmaceutical products to make us better.

All these aspects of modern living and many more are dependent on the chemical industry and there is no doubt that the quality of our lives would be much poorer without them. There is also no doubt that the chemical industry is one which brings the environment/industry dilemma into sharp focus. In making chemicals which contribute so much to enhancing the quality of our lives, the chemical industry cannot avoid having some impact on the environment. (And as private citizens, we also have an impact on the environment by using energy to warm our homes and power our cars and by generating domestic waste and sewage.)

To minimize their effect on the environment, chemical manufacturers need overall management strategies which will include waste minimization as a key feature. Established plants may need bringing up to modern standards and manufacturers wishing to maintain a good environmental reputation may have to strike a balance between the fastest possible improvements which could result in job losses and phased improvements which just meet legal requirements. This chapter will outline how a wholly owned ICI subsidiary, ICI Chemicals & Polymers Limited (ICI C & P), is managing its improvement in environmental performance.

Prior to its recent reorganization, ICI C & P had a turnover exceeding £4 billion and it employed almost 36,000 people in twelve countries. Almost half of its sales were to the UK market and over a third to continental Western Europe, while its products were sold in over 130 countries world-wide. ICI C & P sold more than 800 products and was organized in six business groups ranging from Petrochemicals through Acrylics to Fibres. It was involved with twelve subsidiary and related companies, including European Vinyls Corporation, a very large PVC producer with a turnover of £1 billion/year. ICI C & P's operations were conducted on over thirty sites, the majority of which are in the UK, but some are in continental Western Europe, the USA and Taiwan. One site is located in Kenya.

Systematic Arrangements

To take account of its complex organization and widespread nature, ICI C & P needs simple arrangements for dealing with environmental matters:

- there is a single environmental policy for ICI C & P
- all environmental matters are the responsibility of the line managers concerned
- the ICI C & P Environmental Department advises the Board and management on the implementation of the policy and on standards which ought to be achieved

Soon after the formation of ICI C & P in January 1987, a cross-company survey showed that environmental performance needed improving in some areas and the task is being tackled using systematic arrangements which include:

- the publication of the Environmental Policy
- the establishment of a comprehensive set of environmental standards, together with guidance for achieving the standards
- the auditing of performance against the standards
- measurement of environmental performance and the regular reporting of performance to management and the Board
- the definition and implementation of environmental improvement plans
- the appointment of process and product stewards for co-ordinating our handling of issues relating to our processes and products

Environmental Policy

'Environmental Policy in the ICI Chemicals & Polymers Group' was endorsed by the ICI C & P Board in 1987, and followed ICI thinking closely. It was published widely for implementation by management and the nub of the policy is as follows:

> It is the policy of ICI C & P to manage its activities so as to ensure that they are

acceptable to the community and to reduce adverse effects on the environment to a practicable minimum.

This policy requires that we listen to and act appropriately upon comments from our neighbours. Only by doing so can we continue to earn our notional 'licence to operate' from the communities around our sites.

Research

The policy also requires that we assess the impact of our activities and research means of improvement where necessary. The best current example of this part of the policy in action is the effort that has gone into identifying and then producing 'KLEA', an ozone-benign replacement for refrigerant CFC 12. The first plant in Runcorn has been commissioned and a second is being constructed at St Gabriel in the USA. 'KLEA' is the first of a family of substitutes which ICI is aiming to develop. Other chemicals are being considered for the replacement of CFC 11 in the manufacture of insulating foams. In the meantime, ICI Polyurethanes has already introduced world-leading formulation technology which halves the amount of CFC 11 which is needed in this application.

Moving close to the earth's surface, our research has led to or is offering opportunities to reduce pollution in the lower atmosphere. Two ammonia plants built near Bristol have led to massive reductions in emissions of oxides of nitrogen and sulphur compared to those using conventional technology. Their better energy utilization has also led to a 60 per cent reduction in carbon dioxide emission. If world-wide ammonia production were to be switched to this process, the resulting reduction in nitrogen oxide emission would be equivalent to taking five million cars off the road.

As an additive to allow methanol and ethanol to be used in place of diesel fuel, 'Avocet' is undergoing trials in Los Angeles, Sweden and Switzerland. In Tours, where the French make bio-ethanol from farm surpluses, it is also being used to fuel buses. 'Aquabase' has been developed to replace organic solvents used in car paint spraying and it is currently used by General Motors at Oshawa, Ontario, at Volvo's Gothenberg plant and in Germany by Ford and Volkswagen.

Research in chlorine technology has led to the development of the FM 21 cell which allows production in efficient cells not requiring the use of mercury. Reductions in the heavy metal contamination of liquid effluents are also occurring as a result of the development of 'Syneck'-TAL which can be used to replace chromium compounds in the tanning industry.

Research into biological processes is also proving to be beneficial to the environment. 'Ecosyl' has been developed as a safer alternative to the traditional formic and sulphuric acids used to improve silage fermentation. It also results in up to 40 per cent reduction in the liquid run-off from silage. Many effluents require treatment to remove Biochemical Oxygen Demand (BOD). Where space is at a premium as in Japan, the ICI Deep Shaft process is particularly beneficial. It uses a shaft 50–150 metres deep which is up to ten times deeper than a conventional sewage treatment tank to achieve the necessary oxygenation of the BOD-containing waste. Another development has resulted in an enzyme product which neutralizes potentially dangerous neuro-toxins that are a by-product of water treatment processes using polyacrylamides.

Moving on to solids, ICI has developed 'Biopol', which is both biodegradable and made from a renewable raw material, sugar. Although it is more expensive than the materials conventionally used in plastic bottle production, it is being marketed in Germany, where the level of environmental awareness is high. ICI is also participating in many different recovery and recycling schemes across Europe to reuse PET bottles, for example, as fillings for anoraks, duvets and sleeping bags.

Measurement of Performance and Monthly Reports

The measurement of performance in numerical terms is currently concentrated on the performance of manufacturing sites, with particular attention to the aqueous and atmospheric environments. The introduction of continuous aqueous effluent monitoring at Billingham with alarms set slightly above normal operating conditions (and well below consent levels where these apply) has resulted in tighter control of process conditions (see Figure 9.1). A similar £1 million system has also been installed at Wilton to upgrade previous monitoring arrangements.

At individual sites there is a very large accumulation of environmental information, used for internal control; data reduction is essential for presentation of an assimilable report to the Executive Committee of the Board on a monthly basis. The parameters which have been selected for each separate site are as follows:

- the name of the responsible manager immediately below Board level
- the name of the site
- the number of tests on aqueous discharges and the percentage compliance with consents of tests on aqueous discharges
- the number of tests on atmospheric discharges and the percentage compliance with presumptive limits of tests on atmospheric discharges
- the number of uncontrolled releases greater than one tonne
- the number of incidents reported to the regulatory authorities

In addition to the numerical report, descriptions are given of all incidents reported to the regulatory authorities, of incidents attracting media attention, and of any other incidents meriting description.

Also reported monthly are matters to do with issues; environmental matters associated with products; new legal requirements; progress on environmental improvement plans; and any other matters of considerable environmental significance.

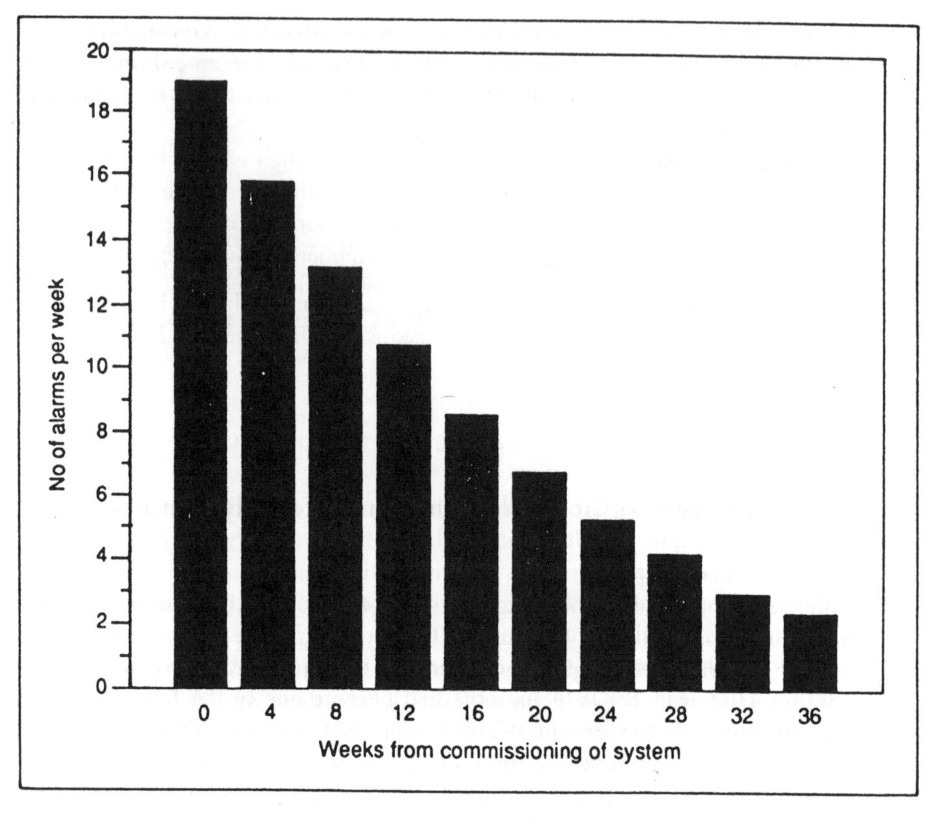

Figure 9.1 Billingham effluent flow – effluent alarm frequency. The introduction of continuous aqueous effluent monitoring has resulted in tighter control of process conditions.

Improvement Plans

ICI C & P adopted a Quality Policy from its inception. One of its principles states:

> We will set ourselves a target of annual improvement in the quality of everything we do.

This principle formalized the long-standing generalized plans for improving environmental performance and each site is now required to have an environmental improvement plan, tailored to its particular needs and including the following features:

- the target of 100 per cent compliance with consents for aqueous discharges and 100 per cent compliance with presumptive limits for atmospheric discharges; and compliance with all other legal obligations, such as waste management and noise regulations

- support for agreed local Environmental Quality Objectives and standards
- a community relations programme including a complaints procedure, arrangements for fostering and improving external appearance, and local information arrangements
- an ecology management plan

To illustrate such plans, two features of the Teesside Operations Environmental Improvement Plan will be described in some detail. These are the River Tees improvement plan and the Teesside ecology management plan. A brief reference will also be made to the reduction in emissions to atmosphere at Billingham.

The River Tees improvement plan

The River Tees rises on Cross Fell, the highest peak in the Pennines, and flows for some 160 km into the North Sea. It has a 40 km estuary and the last 20 km are heavily industrialized.

Industrial Teesside has grown rapidly from the 5,500 people recorded in the 1801 census to a population of about 500,000 today. The discovery of Cleveland ironstone in 1849 boosted steel production and by 1880 shipbuilding had reached 50,000 tons per year. By 1914 the area had become one of the foremost heavy industrial centres in the UK. In 1926 ICI was formed and by 1931 the Teesside chemical industry employed a total of 7,200 people. In the same year, the iron and steel and associated construction industries employed 30,400 people and there were a further 30,000 from these industries who were out of work. Some 2.2 million tons of steel were produced along with 1.2 million tons of pig iron.

The development of these industries and the associated population growth occurred with little regard for their impact on the river. Between 1929 and 1933, the Water Pollution Research Board made a comprehensive survey of the Tees estuary and in each year of observation, large numbers of migratory fish died in the estuary. By 1937 salmon had been virtually eliminated from the Tees. The fish kills were caused mostly by cyanide from the steel industry's coke ovens and an oxygen deficiency arising from the biochemical oxygen demand (BOD) of organic material discharged into the Tees. In 1931, the ratio of the BOD loads from the 49 main raw sewage outfalls, from the coke ovens and from all other industries was 4:2:1 and the coke ovens were discharging 0.8 tons per day of cyanide. The chemical industry grew rapidly and there was recovery in the steel industry, so that in 1945 the chemical and oil industries employed 15,000 people and the steel and engineering industries employed 41,000. By 1970 the former had increased to 34,000 (including 30,000 in ICI) without any significant change in the steel and associated industries; the total BOD load on the river had increased to over 500 tons per day and the cyanide load had risen to 2.2 tons per day.

Gross pollution on this scale clearly could not continue and in October 1968, the Northumbrian River Authority wrote to the Teesside County Borough Council giving notice that:

> The River Authority reaffirm their intention to exercise their powers under the Rivers (Prevention of Pollution) Acts 1951 and 1961 with the object of restoring the wholesomeness of the River Tees.

The local River Authority also urged the Council and the industries concerned to take early and effective action to provide treatment of sewage and trade effluents. With remarkable speed the Council resolved in December 1968 that:

> The necessity to improve conditions on the River Tees be recognized and the need to treat sewage prior to discharge into the river be accepted in principle.

In 1969 the local River Authority and representatives of the major municipal and industrial dischargers agreed that the conditions in the estuary demanded improvement and they decided that 'A reasonable ultimate objective for the condition of the river was such that it would support the passage of migratory fish at all stages of the tide. Initially, however, all that could be expected was a progressive improvement of the river'. The major industrial concerns undertook to make substantial reduction of their trade effluent loads and ICI undertook to reduce its overall 1970 loads by 50 per cent by 1975.

These good intentions were turned into actions. By February 1971 plans for 'Teesside Sewerage and Sewage Disposal' had been prepared and were soon approved in principle by the Department of the Environment. By 1975, ICI had achieved its promised 50 per cent reduction and other manufacturers, notably British Steel Corporation, had also made improvements. ICI continued to make further effluent load reductions, but the river remained polluted.

In 1978 the Northumbrian Water Authority set up a working group to consider the costs and social benefits of alternative pollution control policies. Following its assessment, Northumbrian Water (NW) issued a short consultation paper in April 1980 as a means of obtaining the views of the interested parties on Teesside. The paper outlined the situation as it was in 1979/80, proposed a series of three targets, gave estimates of the associated reductions in pollutants that would be required and the costs to achieve them and then indicated a preliminary view of the way forward. The targets may be summarized as follows:

- Target 1 – Elimination of visual and smell nuisances
- Target 2 – The passage of young salmon to the sea in the spring and the return of adult salmon for several months of the year
- Target 3 – The passage of migratory fish at all times

Although no dates were set for targets 2 and 3, it was proposed that target 1 should be achieved in two stages: the elimination of smell by 1985 and visual nuisance by 1989. In addition to capital already spent, it was expected that the achievement of target 1 would cost in 1980 pounds a further £30 million, that progress from target 1 to 2 would cost more than £50 million, and that progress from target 2 to 3 would cost more than £150 million.

Thirty-six organizations and individuals responded to the consultation paper.

In late 1980, NW held meetings with local authorities, Members of Parliament, representatives from recreational and fishing organizations, and industrial management and representatives of workers in industry. The outcome was that the proposed dates were accepted, and it was hoped that the targets could be achieved earlier.

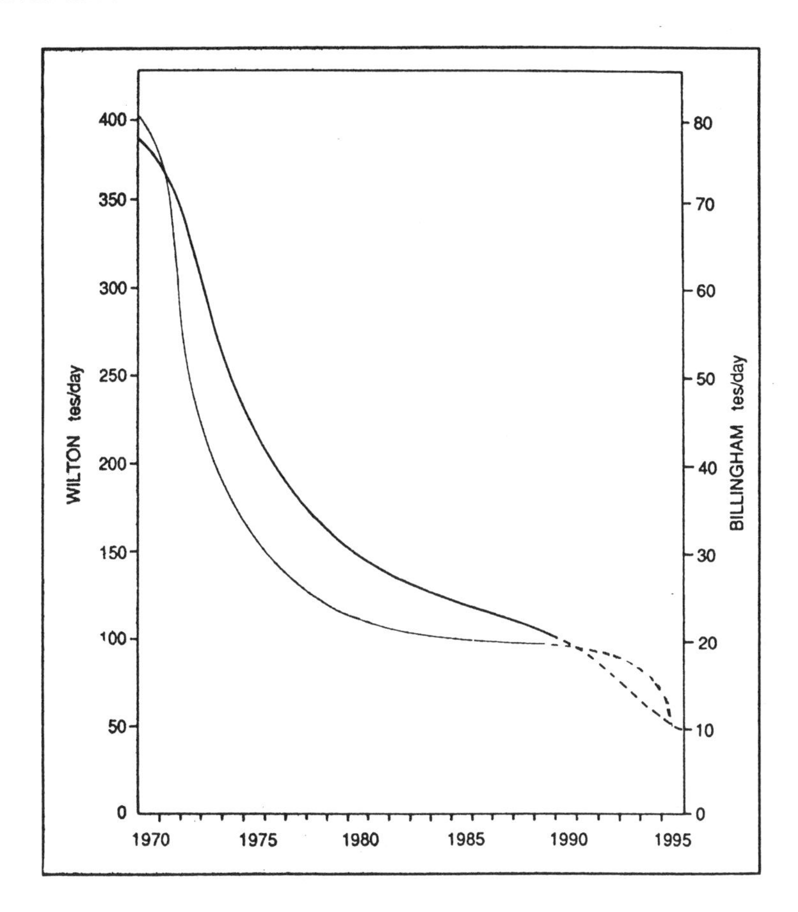

Figure 9.2 Overall BOD loadings in discharges from Wilton and Billingham are now about a quarter of those in 1970 and further reductions are planned.

Improvements in water quality in ICI have been achieved by replacing older industrial plants with plants using up-to-date, cleaner technology, and by adding effluent abatement processes. For example, in 1988 an ammonium sulphate plant was built at a cost of £6 million to remove acidic effluent. This investment was made solely for environmental reasons and it will not show an economic return. Over the past few years ICI has spent a total of about £15 million directly on abatement measures and considerably more indirectly through expenditure on cleaner processes. For example, the Terephthalic Acid plant at Wilton which cost £80 million has a BOD load per ton of product which is a seventh of that from the previous process. Overall BOD loadings are now about a quarter of those in 1970 and further reductions are planned (see Figure 9.2).

Improvements in sewerage together with the introduction of primary sewage treatment have cost NW about £30 million and led in 1991 to the achievement of target 1. Although the river is still polluted, it is no longer toxic to fish and they can now pass through the estuary when there is sufficient oxygen in winter or in periods of high flows of fresh water. The National Rivers Authority (Northumbria Region) report that numbers of young wild salmon are being found in the upper reaches of the Tees, and there are once again reports of rod-caught salmon. In the estuary itself, life is returning. For instance, since 1979 there has been a general increase in the diversity of benthic fauna and their abundance has also increased (see Figure 9.3).

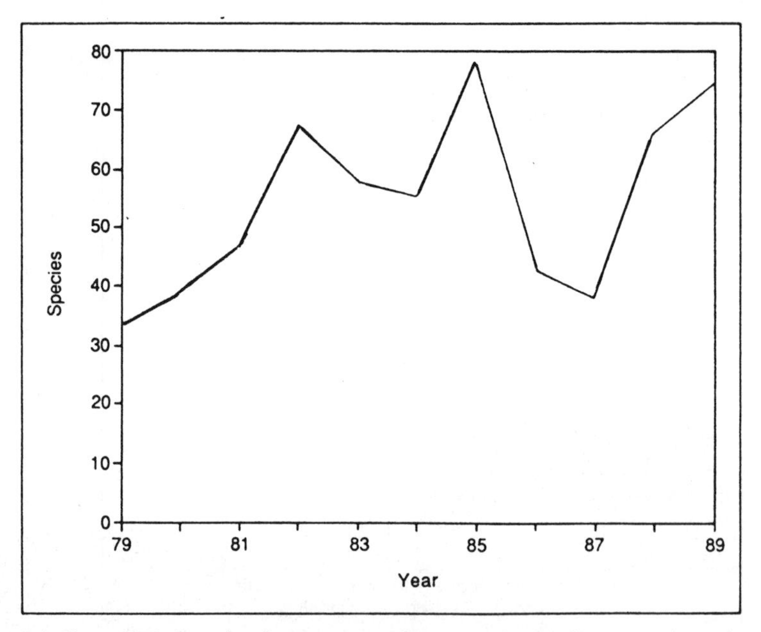

Figure 9.3 Since 1979 there has been a general increase in the diversity of Benthic fauna in the Tees and their abundance has also increased.

Forward planning in ICI is based on the assumption that target 2 ought to be reached no later than 1995. In order to achieve its share of this improvement, it will be necessary for ICI C & P to make significant further reductions in the BOD load from both Wilton and Billingham and to reduce the ammonia content of the Billingham effluent. Several projects will be required. A £5 million reed bed treatment process has already been implemented and a £66 million plant to recover sulphuric acid from the acidic ammonium sulphate effluent arising from the production of methyl methacrylate is being constructed. Overall it is expected that ICI C & P will have to spend well over £100 million in order to achieve its share of the further improvements necessary to achieve the seasonal passage of migratory fish.

Emissions to atmosphere

At Billingham in the late 1960s and early 1970s, the levels of dust/smoke and sulphur dioxide emissions were reduced by 98 per cent and 97 per cent respectively. These reductions came about mainly as a result of the building of new, cleaner sulphuric acid plants and the replacement of old coal-burning equipment with North Sea gas-fired units. These changes had a dramatic effect on the atmosphere at Billingham, as the photographs taken in 1954 and 1984 clearly show (see Figures 9.4 and 9.5). Together with the changes made by other industries and the switch to smokeless fuels in our homes, these changes have led to ground-level concentrations of smoke and sulphur dioxide which are well below the long-term aims of the relevant EC Directive.

Figure 9.4 How it was – view of site in 1954: site enveloped in steamy plumes and industrial aerosol mist.

In the 1970s and early 1980s, old nitric acid plants were closed or modernized and new plants were built operating to very high standards. Although Billingham had the highest concentration of nitric-acid-producing capacity in Europe – perhaps in the world – downwind concentrations of nitrogen dioxide were significantly below the levels found in city centres. For example, the continuous monitors installed by the Department of Environment in compliance with the EC Nitrogen Dioxide Directive showed means and 98 percentile levels of 23 and

63 ppb at Billingham compared with 38 and 85 ppb in central London in 1987. Since that time, fertilizer production at Billingham has reduced with consequent reductions in nitrogen oxide emissions.

Figure 9.5 How it is – view of site in 1984: only cooling tower plumes still visible.

Teesside ecology management

During the industrial development of Teesside, much natural habitat was lost. For example, the intertidal mudflats at the Tees estuary have been reduced from 2400 ha in the late 1800s to about 300 ha in 1987. Developers paid little regard to environmental matters until 1969 when the need for improvement became apparent to all parties. Plans for pollution reduction were developed and at the same time ICI began to give active support to positive nature conservancy.

In 1969, ICI leased the north-eastern section of Cowpen Marsh to the RSPB. This 85 ha Site of Special Scientific Interest is now managed by the Cleveland Wildlife Trust (CWT). In addition to its birds, the area is important for its saltmarsh flora which are the best for a hundred miles to the north and south. In recent years water levels on the Marsh have tended to fall while rising on the nearby ICI farm. This problem has been overcome to the common advantage of both by the installation of pumps and sluices.

ICI has leased Lazenby Bank (100 ha) to Cleveland County Council. The Council manages the site and ICI has retained representation on the management board. Part of the site has free public access, but access to the ecologically more fragile woodland areas is restricted to special interest groups escorted by the warden. More than 40 species of breeding birds and a wide diversity of invertebrates have been recorded in this area, which is also rich in industrial archaeological remains from old ironstone workings.

Norton Bottoms has also been leased to the County and assistance has been given to support a warden. This is an area of pools and marshland in the Billingham Beck valley adjacent to a conservation area already managed by the County.

Gravel Hole, which has been managed by the Cleveland Nature Conservation Trust (now CWT) for a number of years, has now been transferred to CWT ownership. This old quarry is a wildflower and butterfly reserve.

ICI has declared Saltholme Pools to be a reserve. These pools were formed as a result of uncontrolled brine extraction long before ICI came to Teesside. They have a remarkable history of attracting rare birds, especially during the migration seasons. These birds attract bird-watchers from all over the country, sometimes in very large numbers in the case of great rarities like the long-toed stint which had its first UK (and second European) sighting in 1983.

Saline conditions in other parts of the brinefield lead to prolific populations of brine flies which provide food for a very wide diversity of bird life including many rare visitors. Roads in the brinefields are deliberately left unsurfaced and, together with the gravelled areas around the well heads, are much used by ringed plover. With up to 32 pairs holding territories, the brinefield, to which public access is not allowed, is one of the most important east coast sites for ringed plover.

Other initiatives on ICI land include the building of artificial nesting tunnels in an old fly ash bank, to encourage migrating sand-martins to stay and breed alongside kingfishers, which have also bred in the bank.

MSC teams have also planted trees on an old waste disposal site. Much of this site has re-vegetated naturally and with sympathetic management has become a fine wildlife area.

ICI C & P is also involved in similar collaborative improvement programmes on other sites. On Merseyside, for instance, we are actively supporting the Mersey Basin Campaign which aims to upgrade the quality of water throughout the entire Mersey catchment area, and to deal with bankside dereliction. This involves an extensive programme of work to improve the quality of our own discharges and support for Voluntary Sector activities associated with the Campaign.

Following the establishment of the Farming and Wildlife Advisory Groups, the Nature Conservancy Council (NCC) suggested that similar groups should be set up with industry. ICI strongly supported the proposal and, with the support of several other companies in Cleveland, the NCC and the County Council, the UK's first Industry Nature Conservation Association (INCA) has now been established in Cleveland. It is hoped that others will be formed and already there is a similar move on Merseyside.

The aims of INCA may be summarized as follows:

- To promote the importance of adopting the best nature conservation practices
- To promote public awareness of instances where good nature conservation practices have been implemented
- To assist at the interface between industrial and nature conservation interests
- To make available expertise in nature conservation
- To help industrial staff to be better informed about the relevant concerns of conservationists and vice versa

In addition to direct financial support, ICI C & P has released its Senior Ecology Adviser on a part-time basis as a resource for INCA.

Already INCA has worked on five projects for ICI C & P. These include a study on the disposal of dirty salt water which arose during the drilling of a new brine well and the preparation of a wall chart for schools which outlines the ecology of the Tees estuary.

Product and Process Stewardship and Issue Management

Product stewardship

In typical chemical plants, over 95 per cent of the raw materials turn up in the product which is transported to the customer. Only a very small fraction of the production intermediates, by-products and raw materials will be released to the environment at the place of manufacture. It is, therefore, important to be sure that the product will be handled safely, healthily and environmentally acceptably by the customer. There must, therefore, be a responsible and systematic approach to product effects after the product leaves the manufacturing site. The duty of a product steward is to manage these matters. He must be knowledgeable about the properties of the product; he must keep up to date with new information from sources world-wide and he must be aware of legislation in all the countries where the product is sold. He must make sure that appropriate information is transferred with the product to all customers. He may, on occasions, decide that a prospective customer will not be supplied; for instance, the customer may not have safe handling equipment for a hazardous substance. The product steward must be concerned about conditions of storage and quantities stored, and about measures taken to ensure safe storage. He will seek reductions of hazardous material storage and transport. A repeat of the Sandoz disaster must be the product steward's nightmare and he must ensure it does not, in fact, occur. He may commission research work, for instance, to determine the effect of the product in the natural aqueous environment. He will be concerned about reducing the inventory and the transport of hazardous substances. He will assess the options for the recycling of a product at the end of its useful life. He will promote recycling where it is practicable to introduce this practice.

Process stewardship

A process steward will be concerned with the technology and effects of a chemical production process. He will be knowledgeable about the safety, health and environmental hazards of the process and what arrangements must be made to reduce risks to a tolerable level. He will collect information about the process world-wide, and he will have in place information exchange arrangements on safety, health and environmental matters. He will keep any licensees of the process up to date, and he will be concerned with process improvement. He will also be concerned about reducing the inventory of hazardous substances and about eliminating hazardous raw materials and intermediates by changing the process route. He will pay particular attention to waste minimization where practicable.

Issue management

An issue is defined as a substantial topic which is likely to be long-lasting and of considerable current or potential public interest. This definition is not intended to be exclusive, but indicative. Examples of public environmental issues are:

- nitrates in water
- CFCs and the ozone layer
- the pollution of the River Tees

The duties of an issue manager in ICI C & P will depend on the particular issue, and the 'age' of the issue and its relevance to the company; the duties will include:

- co-ordination of company activity on the issue
- management of public relations aspects
- communication of information within the company

The above three roles are intended to systematize the relationship between the product, the process and the environment. Work continues to refine the systematic arrangements.

Training

A key feature of the management of the improvement of our environmental performance is to ensure that all employees are properly trained and fully aware of their respective roles in the overall improvement process. Even those who cannot affect the actual performance can play their part in enhancing the company's reputation by being able to respond to questions from a knowledge of what has been done, what is in progress and the resulting environmental benefits. Information of this nature is shared in several ways, including regular work group briefings, in-house bulletins, newspapers and occasional publications, such as *Environmental Issues*.

Within our operating units, there is a need for a greater understanding of the required environmental standards and for high-quality operation and the elimination of pollution incidents. At plant level, operators are made aware of the potential environmental implications of their acts or omissions. More detailed learning for supervisors and managers is available from distance-learning packages developed by ICI in conjunction with Leicester Polytechnic. These packages cover air pollution control, water pollution control and the disposal of waste. They are designed to meet individual plant needs and they have the advantage that the participant is not tied to timetable or classroom. Participants study at their own convenience and also have access to local specialist environmental advisers for additional explanation and detail. They also have access to environmental data contained in their plant dossier.

Training packages for specialist environmental advisers are developed by small teams acquiring particular expertise in the various aspects of pollution control which are relevant to our businesses. These packages are then used to ensure that the other specialist advisers are adequately aware of what needs to be done to meet company requirements. These include a requirement that all its new plants are built to standards that will meet the regulations it can reasonably anticipate in the most environmentally-demanding country in which it operates that process. This will normally require the use of the best environmental practice within the industry.

Final Comment

With increasing public interest in the environment, most chemical manufacturers are paying more attention to the environmental impact of their operations. This chaper has outlined the systematic approach adopted by ICI C & P in managing the improvement in its environmental performance.

Acknowledgement

The author would like to thank R. L. Pocock, Director, Safety, Health and Environment for his help in the preparation of this article.

References

Porter, E. (1973) *Pollution in Four Industrialized Estuaries* – Four case studies undertaken for the Royal Commission on Environmental Pollution.

ECETOC (1988) *Nitrates and drinking water*, Technical Report no. 27, ECETOC, Brussels.

Shillabeer, N. and Tapp, J. F. (1989) Improvements in the benthic fauna of the Tees estuary after a period of reduced pollution loadings, *Marine Pollution Bulletin* Vol. 20, no. 3, pp. 119–23.

10

THE CHEMICAL INDUSTRY AND ENVIRONMENTAL ISSUES

Steve Tombs

Introduction: the 'Resurrection' of Environmentalism in Britain

Testimony to the contemporary significance of environmental issues is all around us: in national newspapers, through television and radio coverage, on supermarket shelves and in bookshops, in the findings of opinion polls, in the vibrancy of environmental movements, in the national recognition of the Green Party, and – perhaps most importantly – in the 'greening' of all main political parties (Coxall and Robins, 1989, p. 470; see also McGrew, 1992).

It is scarcely surprising that manufacturing industries have been particularly subject to criticism in relation to environmental issues. The association between manufacturing industries and the pollution of air, land and water is a strong one, both in popular consciousness and indeed in reality. But of all the manufacturing industries, the chemical industry has been singled out for particular criticism, due largely to its historical image and the range of its impacts.

Why the chemical industry?

Before considering, in the main part of this chapter, how the chemical industry in the UK has responded to such unwelcome attention, it is worth noting possible reasons why it is that environmentalism represents a particular 'threat' to the chemical industry. It is possible to outline three reasons for this.

First, it is important to consider the nature – or key structural features – of the contemporary chemical industry. Most important of all is the relationship between a chemicals sector and manufacturing in general. The chemical industry performs a pivotal role in any advanced industrial economy. Labelled 'industry's industry' (Ilgen, 1983, p. 650), it has been estimated that

> although the chemical industry in western countries accounts for no more than 3–4

> per cent of GNP and employs under 1 per cent of the workforce, it has a strategic impact on perhaps 40 per cent of the economies of those countries.
>
> *(Davis, S., cited in Pettigrew, 1985, p. 54)*

While the chemical industry produces relatively high returns on capital invested *vis-à-vis* other UK manufacturing industries, the UK chemical industry has performed poorly in relation to its counterparts in the USA, Germany and Japan (Pettigrew, 1985, p. 67). The sector is also characterized by both capital intensity and economies of scale in the industry. Related to each are:

1. the predominance of continuous operations – the producing of vast quantities of homogeneous materials, frequently liquids or gases, which can be manufactured, processed and shipped most economically on a large scale;

2. a tendency towards the building of ever larger plants, driven by the economic implications of the 'square cube law' (Chenier, 1986; Clausen III and Mattson, 1978).

Each has important consequences in terms of environmental hazard. In addition, it has been argued that there has occurred a slowing in the rate at which existing chemical plant becomes obsolescent, with older plant being kept in operation for longer (Witcoff and Reuben, 1980, p. 27). Again, there are important hazard implications associated with this trend.

A second general reason why the chemical industry is a key site of environmental concerns is the hardly deniable fact that there is a sense of the 'unknown' about, and almost a sinister association with, the word 'chemicals'. Much to the chagrin of industrialists, it has been suggested that large numbers of the population do not immediately associate this word with the durability of plastics, the visual lucidity of drinking water, the longevity of pre-packed foods or the comfort of modern upholstery. Quite simply, many associate the word with dangers to health, safety and the environment (Irwin *et al.*, 1988). We shall consider later the extent to which these perceptions – or associations – are either 'warranted' or 'irrational'. That they exist, however, does seem to be clear enough.

Thirdly, and related to the previous point, there either emerged, or occurred, during the 1970s and 1980s a series of health, safety and environmental problems or incidents which, with an almost uncanny regularity, have brought the international chemical industry to public attention. Such scares, scandals or disasters include Thalidomide and Opren, asbestos, various pesticides, Agent Orange, Seveso and Bhopal, Love Canal and Times Beach, and various problems associated with food production. Moreover, such phenomena have increased public sensitivity to regular but small-scale accidents, associated with the production, use, storage or transport of hazardous substances. As a consequence, the chemical industry has an unfriendly image, and has become the *bête noire* of environmentalists. Given these three general points, then, it should be no surprise that Jean-Marc Bruel, President of the Rhone-Poulenc Group, has stated:

> More than any other industry, chemistry finds itself at the cutting edge of our relationship with the environment.
>
> *(Bruel, 1990, p. 736; see also Horton, 1990, p. 747)*

More emphatically, and in a sense setting the scene for this chapter, Edgar Woolard, Chair and Chief Executive Officer of DuPont, has stated of the challenge posed to the chemical industry by the resurrection of environmentalism:

> The future of the chemical industry will be directly shaped, and indeed may ultimately be determined by environmental issues.
>
> *(Woolard, 1990, p. 738)*

The strategic responses of the industry to this challenge are, therefore, of key importance.

Environment, Risk and the Image Debate

In conjunction with the resurrection of environmentalism in the UK, the chemical industry has been undergoing an anxious and introspective debate concerning its public image. Changes in public perceptions of the chemical industry have been charted in a series of polls conducted by the MORI organization from 1969 onwards. Liardet, of the Chemical Industries Association (CIA), has noted that after almost ten years in which the favourability ratings of the industry 'changed very little', this rating began to fall quite dramatically towards the end of the 1970s (Liardet, 1991, p. 118). From a favourability rating of about 50 per cent of respondents during the period 1969–79, this started to fall ominously to 41 per cent in 1980 and 38 per cent in 1981 (ibid.). By 1987 such polls were recording that fewer than 33 per cent of respondents had a favourable view of the industry (Lindheim, 1989, p. 491), while in November 1989 just 26 per cent were favourably disposed towards the industry (Liardet, 1991, p. 118). It is important to emphasize that these findings of a poor – and declining – reputation cannot be directly related to 'ignorance' on the part of the general public. While MORI has posited an association between favourability and familiarity, its polls show that between 1981 and 1989 there was recorded a slight increase in public familiarity with the industry, thus raising 'the ugly thought that the public is hearing more about [the chemical industry] and, in the context of a greener world, does not like what it hears' (Liardet, 1991, p. 119). Thus while it is of course important to be cautious about the interpretation of such findings, it is not good enough to argue that such figures derive from 'a vast area of sheer ignorance' on the part of the British public (Roddom, 1991).

Arguments about the unfairness or unjustness of such a perception are one type of response made by representatives of the industry. However, before considering this and other responses, let us develop our examination of the 'image' question, this time more specifically in relation to the chemical industry and the environment. As one might expect, the fact that the industry's public reputation has deteriorated in an era which has witnessed the resurrection of environmentalism is not, of course, coincidental – the two phenomena are, as has already been suggested, directly related.

The MORI polls cited show that the 'general public' is quite clear about what it does not like. Towards the end of the 1980s, 'over 90 per cent were concerned about chemical pollution', with three out of four respondents thinking that the chemical industry did serious damage to the environment, and almost half of these thinking that it 'did not care' about these issues (Lindheim, 1989, p. 491). Similarly, 87 per cent of respondents said they were 'very concerned' about the disposal of chemical waste, while over half were 'very concerned' about plant safety, storage and transport (Liardet, 1991, p. 119). Worryingly for the industry, and somewhat contradicting claims that such opinions are the result of 'ignorance', MORI research findings also showed that the vast majority of persons in AB (professional and managerial) social groups have learnt about the chemical industry via publicity about accidents or pollution (80 per cent and 70 per cent respectively) (Grant *et al.*, 1988, p. 243). Amongst AB men, traditionally the group most favourable to the chemical industry, this favourability was falling, the main explanation being 'a concern with pollution' (ibid., pp. 265–6). Relatedly, then, these same social groups recorded diminishing scores on the question of the 'perceived usefulness of the industry' during the 1980s (ibid., p. 266). According to Grant and his colleagues, such data represent 'a value change ... among the best educated sections of the community', these sections being those 'from which the legislators and regulators are drawn' (ibid., p. 268).

We are now building up a picture, both of how the chemical industry is popularly viewed, and how it sees itself as being viewed. The industry's declining reputation is crucially tied to public perceptions of its role in both acute and long-term damage to public health and the environment. As we have indicated, a continual theme of representatives of the chemical industry is to emphasize the importance of the industry and its products to the economy, the quality of life, and so on. Indeed, as will be considered below, a related response to environmentalism is to argue for the centrality of the chemical industry in solving environmental problems. In projecting the industry as part of the solution rather than the problem, in seeking to 'put into perspective' the role of the chemical industry in causing environmental problems, in comparing this with other industries, and in emphasizing the importance of chemicals to our national economy and to contemporary daily life, the chemical industry appears to be using various techniques of denial or neutralization.

We have also begun to see other elements creeping into the industry's assessment of its current position *vis-à-vis* the public and the environment. The implication that its low public esteem is unfair is rarely far from the surface of much of this rhetoric. Even more important, however, is the notion not just that this reputation or image is unfair, but that the 'public' is ill-equipped to make such judgements.

The Strategic Responses of the Chemical Industry to Environmentalism

Given the conflictual environment in which the chemical industry finds itself, it is not surprising that corporations have embarked upon aggressive marketing strategies aimed at cleaning up this negative image. It is possible to identify five broad strategies adopted by the industry in attempts to negate the surge in environmentalism, namely: the 'therapeutic alliance'; industry association initiatives; process innovation; dialogue over risk acceptability; and, finally, the existence and exercise of (technocratic and financial) power. Each of these strategies will be explored in turn prior to assessing their significance for the industry in the 1990s.

Strategy 1: the 'therapeutic alliance'

The basis of this strategy is that environmental concerns on the part of the public are largely emotional rather than 'rational'. In this sense, it is the strategy that draws most directly and exclusively upon the kinds of claims concerning 'irrationality' that were cited immediately above. In its worst manifestations, this response can simply urge the need for education of workers, consumers, users, and so on. In other words, one element of such a response approach can be classic 'victim blaming'. For example, having claimed that the disasters at Bhopal, Sandoz and Basle 'had their origin in human error', Todd goes on to use the example of the production and use of pesticides to illustrate his belief that remedial attention should be focused less on producers and manufacturers of chemicals, and more on their users: 'What we need is better education and control of users' (Todd, 1987, p. 641). The problem, then, is not with the chemical (or here pesticides) industry. It lies elsewhere – in this case with users, who might be consumers or workers.

Such a view was precisely echoed in a series of glossy leaflets produced by the Chemical Industries Association in 1986. This series highlighted the significance of chemicals for everyday life, noted an apparently good record on environmental, safety and health issues on the part of the industry, and emphasized various commitments on the part of the industry to responsible practices. Significantly, the role of the users of chemicals, particularly consumers, was also emphasized, claiming that when members of the public do receive information concerning safety and health they often do not act upon it, that is, they do not act rationally:

> In the home, people use some very strong, powerful chemicals ... On each packet or bottle there are carefully worded precautions which it is wise to follow. A surprising number of accidents occur through failure to follow this guidance, which the makers prepare with great care.
>
> *(CIA, 1986)*

Thus certain industry spokespersons appear to believe that members of the public, even where they are aware of a particular hazard, nevertheless fail to act rationally in the light of that knowledge. Here we encounter the assumption on the part of industrialists and 'experts' that, even if the public is given the 'facts', it remains 'deficient in proper reasoning powers' (Perrow, 1984, p. 315). As indicated above, these ascriptions of irrationality are also used by the industry to perpetuate 'careless worker' explanations or popular understandings of accidents (Tombs, 1991); indeed, such is the power of this ideological portrayal of the relationship between workers and industrial accidents that it persists in explanations of particular incidents even where there exists significant evidence to the contrary (Pearce and Tombs, 1989). In later sections of this chapter, we shall take up charges of irrationality; prior to this, we shall examine how they are central to chemical industry responses to environmentalism.

An approach somewhat more enlightened than pure 'victim-blaming' has been urged by Lindheim. He 'recognizes' that when dealing with popular risk perceptions, the chemical industry is

> in the realm of the illogical, the emotional, and we must respond with the tools that we have for managing the emotional aspects of the human psyche.
>
> *(Lindheim, 1989, p. 493)*

The solution to the image problem, then, is not based upon 'information and facts', since 'Wisdom suggests that you must deal in emotional arenas with emotional tools' (Lindheim, 1989, p. 494; and see Liardet, 1991, p. 123).

Thus Lindheim argues that the chemical industry needs to build a 'therapeutic alliance' (Lindheim, 1989, pp. 493–4): the industry must express its desire to minimize environmental damage, while at the same time reminding the public of the benefits of chemical products, and their role in protecting the environment; moreover, these efforts must be adequately resourced (ibid., p. 494). This 'therapeutic alliance' is to be formed out of 'an army of spokespeople speaking on its [the chemical industry's] behalf', these spokespeople including farmers, manufacturers, packagers and retailers, and developing to include doctors, nutritionists, academic scientists and even the World Health Organization (ibid., p. 494). Relatedly, the chemical industry should also improve its relations with the media, through regular briefings and a commitment to openness (Liardet, 1991, p. 120).

The 'needs' of the chemical industry, on the basis of this response, are clear. In the words of Frank Popoff, President and Chief Executive Officer of Dow Chemical,

> The industry must dedicate the time and the people, as well as the funds, to take the story of science and its benefits to the public, over the long term.
>
> *(Popoff, 1989, p. 755)*

Thus, through the development of a therapeutic alliance – through repetition of, and thereby increasing familiarity with, a soothing story – the public will be nursed out of its irrational (mis)conceptions *vis-à-vis* the role of the chemical industry as a key contributor to contemporary environmental problems. As noted, this is also important in the related context of accidents and plant safety: by continually representing individual workers (through 'human error') as the causes of accidents, the public should be reassured that managerial and organizational systems for operating plants safely are, barring unforeseen actions on the part of subordinate individuals, largely sound.

Strategy 2: industry association initiatives

While the 'therapeutic alliance' refers to a strategy adopted by certain individual sites and companies, the second main response on the part of the chemical industry to the resurrection of environmentalism has been organized at the sector level. In the latter part of the 1980s, the industry launched several and various trade or industry association initiatives. The most important amongst initiatives developed by the CIA include 'Open Door '86', when over 200,000 people visited chemical plants, the launch of the CIA's 'Chemicals in the Community' programme, and, more recently, the CIA's 'Green Code' and 'Responsible Care' pledges.

The basis of such approaches has been neatly summarized by Dewhurst. Arguing that 'the chemical industry is often too negative about its ability to improve public perceptions', Dewhurst concludes that 'the communication problem [*sic*] can be met and overcome if it is tackled with enthusiasm, energy and commitment' (Dewhurst, 1991; see also Horton, 1990, p. 747; Malpas, 1987, p. 646). Often having been presented as 'self-regulatory' programmes, though in reality proving to be largely major public relations exercises, such approaches do at least represent both a recognition on the part of the industry that it must take some notice of environmental concerns, and an indication of a willingness to provide some information relating to chemicals production (and indeed storage, use, transportation and disposal). However, these approaches remain of strictly limited value. The CIA's 'Responsible Care Programme', for example, is based around an approach which seeks to reassure publics through the provision of 'relevant' information, and through improving the reputation of the industry (see Morrow, 1989, p. 280). Similar initiatives have emerged in continental Europe, both within and without the European trade association CEFIC (see Henry, 1988), in the USA and Canada, and even at the international level, in the form of the UNEP APELL programme (see United Nations Environment Programme, 1988).

All such initiatives have typically revolved around information giving and persuasion, aiming at generally 'allaying fears' (Popoff, 1989, p. 756). They have involved the development of fact sheets for public consumption, the attempts to use workers as 'ambassadors' for the industry, the exploitation of 'communications opportunities', and so on. Fundamentally, such initiatives begin from the premise that publics need to be disavowed of their unjustifiably critical attitude to the activities of the chemical industry.

Moreover, while the impetus for such developments partly derives from 'political fallout' from disasters such as Sandoz and Bhopal (Bowman and Kunreuther, 1988), such impetus is more importantly related to the threat of impending legislation enacted by national or even supranational authorities (see, for example, Stover, 1985). The initiatives, then, are not only public relations exercises in the context of the general public; they are also directed at governments and regulators. Thus they seek to provide evidence, at a highly generalized level, of a willingness to self-regulate and act responsibly (see Pearce and Tombs, 1991b). In other words, they are strictly tactical, and have certain defining characteristics: developed by the chemical industry itself, they are 'pre-emptive' and under the control of the industry; they do not seek to facilitate the voicing of concerns of various interested publics; and they make a virtue of necessity, obscuring the fact that any 'responsible' practices are likely to have been forced upon industry, by legislation and the fear of legislation (Pearce and Tombs, 1991b).

Strategy 3: process innovation

A further response on the part of the chemical industry to the resurrection of environmentalism has been to invest in new – and cleaner – process technologies, as well as increasing investment in pollution abatement hardware. Despite an element of 'technical fix', such strategies are generally to be welcomed since they constitute a more positive response to environmental concerns.

As noted earlier, the UK chemical industry is highly capital-intensive, with huge start-up costs and high levels of research and development expenditure, and is a relatively poor player in a highly competitive global market. Given these points, additional investment in pollution abatement equipment and new technologies in general adds to already high costs of producing and competing in the chemicals sector. Pollution abatement costs are now becoming significant in affecting the competitiveness of western industry in the face of stiff competition from the 'developing' world, where higher pollution levels are tolerated as the price for economic prosperity.

A recent report on the costs of pollution abatement/control equipment found that the UK chemical industry spends more on environmental protection than any other branch of industry in Britain. It was noted that in 1988 the industry spent almost £400 million on pollution control, out of a total expenditure by all industrial sectors (public and private) of £3,800 million, and compares with the second highest spenders, the food industry, at just over £300 million (Ecotec, 1989, pp. 23, 35). In 1989 the industry spent over £600 million on capital 'that can realistically be categorized as aimed at environmental improvement' (Liardet, 1991, p. 119). Moreover, expenditure at such levels has been made in the context of an industry which has been hit severely by recessions in the early and late 1980s. More recently, it has been reported that while 'recession has hit the chemical industry particularly hard', leading to a general 'slashing' of investment, chemical companies are spending an increasing proportion of that investment on pollution control. Thus such spending is expected to grow from a current 11 per cent across the industry to almost 25 per cent in the next two years (Cowe, 1991).

Such responses are necessary, if overdue. Bruel has noted the extent to which the resurrection of environmentalism revealed gaps in scientific knowledge, questioned some of its assumptions, and even uncovered examples of negligence within the industry (Bruel, 1990, p. 734). Having been forced into action by environmental pressures, the chemical industry cannot return to its 'old ways', and must 'integrate environmentally friendly practices into our technical and operational objectives' (ibid.)

We should note, however, the way in which such a recognition can lead to a simple 'technical fix' position, one which attempts to insulate further the chemicals industries from external criticism and indeed regulation:

> The answer to the environmental challenge lies at the heart of our industry, in its technology. Developing clean technology is clearly the first step in our response.
>
> *(Bruel, 1990, p. 734)*

This position has been articulated most crudely by Liardet:

> The ability of the industry to innovate towards a cleaner world – and it is only the industry that can do this – depends upon brute profit.
>
> *(Liardet, 1991, p. 120)*

Here, then, we have a statement that clearly emphasizes the fact that such a response is aimed not just at responding to the resurrection of environmentalism, but also at pre-empting stricter, external regulations by demonstrating that the industry can, and must, self-regulate. The implication here, elsewhere articulated quite explicitly, is that external regulation can be misdirected, unnecessarily burdensome, bureaucratic, and involve wasteful compliance expenditures, thus damaging profitability and undermining the ability of the industry to put its own house in order (see Pearce and Tombs, 1990, 1991a).

It is from such a position that the US Chemical Manufacturers Association (CMA) has appropriated an old radical slogan of the 1960s – namely, that one is either part of the solution or part of the problem. As Pearce (1990) has argued, this has been used directly in recent advertising campaigns which ask audiences to trust the industry's 'technofix' to modern environmental problems (and see also di Meana, 1990, p. 744; Trowbridge, 1987, p. 647). Moreover, as the very existence of this advertising campaign itself indicates, it is not enough for new technologies, processes and products – either contributing directly to the protection of the environment or reducing damage to the environment by the industry – 'simply' to be developed. The public must, of course, be made aware of their development, and thus of the role being played by the chemical industry in protecting the environment. Bruel argues that examples of new technologies and corporate improvements 'must work to our advantage and be brought to the attention of the public, which is clearly not up to date or *au fait* with the state of the chemical industry' (Bruel, 1990, p. 736).

There are numerous specific examples of clean technologies being developed by chemicals companies in the UK (see, for example, Bruel, 1990, p. 734–5; and Woolard, 1990, p. 739). It is important to note, however, that while such efforts are represented by the chemicals industries as evidence of their ability to self-regulate, such improvements have only been made as a result of external pressures, many of which are related to the resurrection of environmentalism. It is worth noting, for example, the present 'battle' between ICI and DuPont in their efforts to become the first company to develop a commercial plant for producing a non-ozone-depleting CFC alternative (*Chemistry & Industry*, 1989;

1990a). These companies were, of course, among those that had for years, and even until very recently, denied mooted links between the use of CFCs and depletion of the ozone layer. DuPont was the major producer of CFCs and even in 1988 was opposing their regulation, branding as 'irresponsible' those scientists who warned of their danger to the environment (Pearce and Tombs, 1991b).

Finally, it is worth noting the extent to which such process and technological innovations are linked by key industry representatives to Total Quality Management (TQM) (on Total Quality Management and the chemical industry, see CIA/BSI, 1987; CIA, 1989; Francis, 1990). For example,

> The myth that environmentally safe operations must always result in added costs is false. We need to communicate that to our operating people, to our engineers, and to our customers.
>
> *(Woolard, 1990, p. 739)*

Similarly, Horton has noted that 'innovative technology' can 'make improving the environment pay' (Horton, 1990, p. 749), adding that the principles of total quality management, if properly applied, could allow companies to become both more competitive and environmentally friendly (ibid., pp. 748–9). In accordance with the philosophy of TQM, then, we might expect more leading companies to follow the example of ICI and move from specific process and technological innovation to making public and far-reaching commitments to environmental protection. Thus ICI recently announced plans to 'halve waste emissions in five years' (*Chemistry & Industry*, 1990b).

Such specific commitments are crucial and to be applauded. However, equally crucial are mechanisms by which ICI, in this example, could be judged in practice against their rhetoric. One means of this could be via environmental auditing. Interestingly, on this question, EC plans to move towards integrated environmental control of industrial facilities will include some form of 'environmental auditing'. This will require both the development of an equivalent to a CIMAH 'Safety Case' in respect of the environment, as well as the provision by companies of 'an environmental statement, based on the results of the "audit", to be made available to the public' (di Meana, 1990, p. 746). Given this, it is unfortunate, then, that the chemical industry has been fighting to keep results of any EC-inspired mandatory environmental auditing secret (see Rose, 1990). Indeed, more generally, the UK chemicals industries argued against the inclusion of a clause requiring the development of a computerized database of emissions and leaks from plants in the new Environmental Protection Act, as exists in the USA in the form of the Toxic Release Inventory. This opposition was voiced (successfully) despite a recognition by John Colemen, Environmental Affairs Manager at ICI, that the public are 'better served' in terms of information and therefore potential action by such an information system (*Dispatches*, 1991)

Such positions rather undermine claims to greater openness on the part of the industry. They also highlight the limited extent to which the industry may be able to enter into genuine dialogue – if all interested parties do not have access to relevant environmental information, then communication remains highly one-sided.

Strategy 4: dialogue and 'acceptable risk'

This fourth strategic response by the chemical industry to the resurrection of environmentalism has two closely related elements. Firstly, it is recognized that the tasks of managing the industry in a way that minimizes hazards associated with safety, health and the environment requires fundamental and thoroughgoing changes in forms of management. Secondly, those advocating this strategy accept that such an approach also requires that companies and their managements engage in dialogue with at least some of their 'opponents' or critics in the context of safety, health and the environment.

The phrase 'safety, health and the environment' rather than simply 'the environment' is used quite deliberately. For this fourth response, with its tendency to approach management in a holistic fashion, recognizes that these three areas can only be protected adequately if protection of each is seen as closely related, and, indeed, integrated with other management goals such as efficiency and profitability. Clearly, then, this response also adopts the TQM elements of strategies centred around process and product innovation discussed above. But it is in the recognition of entering into dialogue with groups outside management – those with formal executive authority within and over chemicals plants – that this strategy becomes simultaneously distinctive and more radical.

In this context, a key question often addressed by leading executives of chemicals companies is that of acceptable risk: what is acceptable and, more importantly, who decides this? The ways in which companies seek to answer this last question is a useful indicator of the extent to which they have embraced or adopted this fourth strategic response.

Malpas has considered a variety of possible answers, beginning with the consideration that consumers may decide what is 'acceptable'. This possibility is rejected, however, since he argues that the means by which they would exercise such choices and decisions would be through the market, in which consumers are 'in most cases ... primarily interested in price and performance', and thus 'are not always willing to accept the costs of environmental protection from the manufacturing and distribution operations' (Malpas, 1987, p. 645). Such a position is undermined by recent social science research on the responses of workers and consumers to information as opposed to exhortation policies in their shopping and working habits (Viscusi *et al.*, 1987). Notwithstanding this point, Malpas continues, and having also rejected the possibility that government can or should decide what constitutes acceptability, he is quickly led to the conclusion that since 'nobody is perfectly placed to decide',

> The best solution is a process of consultation and co-operation which will allow a sensible consensus view to be formed – much as the system works today.
>
> *(Malpas, 1987, p. 645)*

Now, we can agree that consultation and co-operation are crucial in answering the question, what is an acceptable risk; but, *contra* Malpas, this is definitely not 'much as the system works today'. For within this 'system' there are no mechanisms which allow an input on the part of workers, local community groups and wider publics in this decision-making process – and it is instructive that the possibility of such groups having a legitimate or useful role to play in this decision-making process is not even considered by Malpas. Thus when he adds that 'Ultimately, society decides where the cost/benefit and risk/benefit balances are to be struck' (Malpas, 1987, p. 646), he operates with an extremely restricted view of who or what groups constitute 'society' (see also Trowbridge, 1987, pp. 648–9).

A more progressive consideration of the question concerning participation over risk acceptability is found in the arguments of Woolard. Recognizing the past errors, insularity and lack of responsiveness on the part of the chemical industry in the context of environmental protection and public expectations, he argues the need for the development of a 'corporate environmentalism'; this is defined as 'an attitude and performance commitment that places corporate environmental stewardship in line with public desires and expectations' (Woolard, 1990, p. 738). Expanding upon a potentially radical and highly progressive approach, Woolard proclaims DuPont's intentions to improve its environmental act (ibid., pp. 738–9), emphasizing that one element of this will be to meet and discuss with 'responsible' or 'mainstream' environmentalists. Of such groups, he states that 'We may not agree on every issue, but we agree on the big picture', since they share the goal of a 'sustainable world' (ibid., pp. 739–40).

Here, then, we have the senior executive of one of the largest chemicals multinationals in the world publicly committing that corporation to engaging in a genuine dialogue and co-operative efforts in order to further environmental protection. Of course, such words may not be translated into practice, but such public commitments are important, since they can then either be monitored, or at least serve to strengthen the arguments of 'outside' groups that there must be developed (legal) mechanisms through which such monitoring can be achieved.

That being said, the limits to such an approach are quickly highlighted in a reference to 'radical environmentalists' (Woolard, 1990, p. 740). These limits to dialogue have also been clearly expressed by the Director of Environmental Communications at Monsanto, who has pointed to a fundamental incompatibility between the chemical industry and certain of its critics:

> You are not going to get anywhere sitting down with the Greens, and we are not going to get anywhere sitting down with Greenpeace.
>
> *(Bishop, 1985, p. 175)*

Here, then, we have a clear expression of which groups are to be legitimately involved in consultations, let alone in formalized decision-making processes, around what constitutes 'acceptability'. However, it is important to add *some* qualifications to these rather negative comments concerning the nature of the dialogue envisaged by the chemical industry. Woolard has also indicated that DuPont are tentatively approaching 'representatives of more radical environmental groups like Greenpeace' in an explicit attempt to widen the limits of dialogue. Moreover, in the UK, there are some traces of a *rapprochement* between the chemical industry and some of its more 'radical' environmental critics. These signs are impressionistic, patchy, and hardly a basis for generalization. Yet they perhaps represent the first steps in a shift from a purely 'technocratic' approach to managing environmental protection.

Strategy 5: the existence and exercise of power

A fifth strategy not explicitly considered thus far in the context of the resurrection of environmentalism has been the existence of power within, and its exercise by, corporations in the chemical industry. The exercise of power cuts across, but is distinct from, those strategies discussed above. With the 'un-politics' (Crenson, 1971) or the dormant and passive phases (Blowers, 1984, pp. 47–50) of environmental issues having passed, the chemical industry now finds the environmental effects of its activities subject to varying levels of scrutiny, protest and opposition, and the power which exists or resides within the industry becomes more manifest as it is exercised. There are four main bases for the power of corporations within the chemical industry.

1. The pivotal role of the chemical industry in any advanced national economy leads us to expect governments of most political persuasions to lend support to this industry when under attack from critics.

2. There is financial power deriving from the multinational nature of the corporations which dominate the industry. This has two aspects. Firstly, these corporations have access to quite massive resources used to mobilize opposition against critics and to pre-empt regulatory efforts. Secondly, the multinational character of many of these corporations means that the threat of a relocation of their activities and future investments is a real one. This reality is related to the existence of states in which ruling élites are ready to accept the introduction of hazardous technologies, the social and environmental effects of which are to be borne by (sections of) their population (Ives, 1985).

3. The fact that the chemical industry represents almost the paradigmatic scientific-technological industry is significant in terms of technocratic power, again in two respects. First, this allows corporations to deny legitimacy to their critics on the basis that they lack the information and understanding necessary to intervene in what are represented as essentially technical issues. Secondly, and relatedly, the highly competitive nature of this scientific-technological industry provides a rationale in terms of 'commercial secrecy' for maintaining strict control over information.

4. While the industry is a huge one, it is important not to confuse its physical and human scales. The global industry is overseen by a relatively small number of key players, constituting a technocracy, which, by means of organized channels, can organize opposition to environmentalist and governmental or regulatory critics (Grant *et al.*, 1988, p. 7).

Massive power therefore *exists* with corporations within the chemical industry; and in the context of conflict over the environment, this power has been used in various ways. As indicated above, corporations have sought both to deny the existence of environmental problems through attempts to manipulate various publics, and have also used their financial power quite bluntly and crudely, either through seeking to co-opt individuals from oppositional movements or by silencing such opponents through intimidation and the threat of legal action (Smith, 1990, p. 24). Corporations have also sought to limit access to information which might provide critics with greater tools and thus lead to moves towards some 'equalization' of power, through rendering communication between corporations and their critics less 'distorted' (Habermas, 1970). This latter point indicates the significance of even the tentative steps towards dialogue considered as strategy 4. Yet it is important not to represent such shifts as deriving from altruism on the part of those corporations involved. Strategy 4 represents a recognition of the need to respond more positively to the emerging power of the industry's opponents, this related to 'a shift in values throughout the advanced industrialized countries' (Blowers, 1984, p. 49). Thus, despite the massive financial and technocratic power that resides within the chemical industry, it remains the case that all forms of power are relational and involve some forms of negotiation, so that where there is power there is resistance (Foucault, 1980, p. 142).

If the power of corporations does not exempt them from making concessions, nor is this power to be seen in a monolithic or reductionist sense. The attempts by larger corporations to pre-empt stricter regulation at the national and international levels, through a technocratic representation of environmental issues as soluble only by the industry itself, indicates the lack of any unified position on the part of the capital interests involved in this sector. These larger corporations have sought to define the problems of safety, health and the environment which have come to be associated with the industry as problems of the less socially responsible element among them, these usually being represented as smaller (Pearce, 1990). This rhetoric is based upon elements of truth – larger corporations in the industry are spending relatively more on pollution control than their smallest counterparts (Ecotec, 1989, p. 32). But this rhetoric has also proved useful since it has coincided with the interests of the mainly right-wing governments that held formal political power in much of the western industrialized world in the 1980s; thus distinctions between responsible and irresponsible actors have provided various regulatory agencies, under material and ideological attacks, with justifications for a selective 'concentration' of enforcement resources.

Larger corporations have also urged that regulations currently in existence are strictly and consistently enforced. This may seem surprising – yet it must be borne in mind that such statements, usually made by leading members of larger chemical companies or industry associations, are actually demands for the equalization of competition within (Marx, 1976) and across (Leonard, 1988) national boundaries. In this respect, we can perceive conflicting interests for different sectors of capital.

Thus it is crucial to recognize the existence and exercise of power in the context of the chemical industry and environmentalism. Equally, however, it is of both theoretical and practical importance to avoid cruder forms of analysis which see the power of capital in a reductionist sense.

Conclusions: Publics, the Chemical Industry and the Environment – Beyond 'Irrationality'?

It has not been intended to give the impression in this chapter that the various strategies are five different *types* of response from which companies will choose. Clearly, companies will – and do – adopt strategies based upon a combination of each. Moreover, as should have become clear, there are some common elements across the different policies. In this sense, strategies 1–4 are best considered as progressively more all-encompassing and adequate strategies. Strategy 5 is somewhat different, since it encompasses a key resource and feature of chemical companies through which strategies 1–4 might be acted upon. What is important within the context of this discussion is which strategy is emphasized or predominates in any particular company's approach to the environmental issue.

Moreover, there is further break or discontinuity in the approaches outlined, and this is best understood in relation to the ascribed irrationality on the part of environmental critics, publics, workers, and so on. The strategies of the 'therapeutic alliance' and the 'industry association initiatives' each see public fears concerning the chemical industry and the environment as fundamentally irrational, and thus requiring initiatives at the level of rhetoric or discourse, rather than practice. At the heart of each of these approaches, then, are the notions that the chemical industry has been unfairly represented and is ill-understood, so that the major tasks of the company or industry are of a presentational, public relations nature. The third and fourth strategies are rather different. They retain elements of these notions, yet also accept that an adequate response to environmentalism requires changes at the level of practice. Of course, while one element of strategy 3 is that the publicizing of changes to working practices, processes and technologies will allay public fears, it is recognized that such changes must move beyond rhetoric. A further important feature of this third strategy is the assumption that answers to the problems set the industry by environmentalism and environmental critics will largely emerge from within the chemical industry itself, from amongst the *legitimate participants* in the problem-solving process.

The fourth strategy also accepts the need for tangible changes and improvements, but recognizes the need to engage in dialogue with opponents and groups which the other strategies do not recognize as legitimate participants. In other words, this strategy accepts that inputs to problem-solving may originate from 'outside' the industry. Yet even here, 'irrationality' intrudes as a quality attributed, not to all critics nor even to active opponents of the industry, but 'merely' to certain 'irresponsible' or 'political' groups or individuals.

These points having been made, it should therefore be clear that strategies 1–4 are largely possible through different forms of power, the fifth strategy. To the extent that companies embrace more progressive strategies, namely 3 and 4, then they must be seen as shifting – however tentatively – from a purely technocratic mode of existence and operation. Those companies primarily or predominantly adopting strategies 1 and 2 will fail seriously to address environmental problems, while companies approaching these issues through strategies 3 and/or 4 will have greater success, despite the fact that these approaches themselves remain flawed. Let us emphasize why this is the case. Strategies 1 and 2 largely fail to treat environmental concerns, and in particular those voicing these, as 'rational'; according to these strategies, the main tasks for the chemical industry are presentational ones. Strategies 3 and 4 have the merit of treating *some* environmental concerns – or rather, concerns voiced by some groups, since it is often the source of speech which is crucial (Tombs, 1990) – as legitimate and rational ones, and thus are more likely to adopt significant changes at the level of practice rather than simply at the level of public discourse.

In addition to the arguments made in this chapter about the problems associated with ascribing 'irrationality' to various groups or individuals outside the industry, there is one final reason why such ascriptions are increasingly difficult to sustain, namely, the industry's record on environmental, safety and health issues. In the context of both the environment and health, the industry has traditionally adopted techniques of denial which have resulted in what David Gee has referred to as 'generation games' (Gee, 1987). In other words, evidence from workers, community groups and 'alternative' experts has often been ignored until the tangible evidence of environmental destruction, ill-health or deaths has become overwhelming.

What we have been considering in this chapter, then, largely relates to the inadequacy of the concept of '(ir)rationality' in relation to risk which is utilized by the chemical industry. Thus, common (industry) uses and understandings of rationality, expressed as risk assessment and risk perception, need to be redefined to take public experiences into account (Sheldon and Smith, 1991). Crucial to such redefinitions must be the transcendence of the 'objective risk – public hysteria' framework (Wynne, 1989a, p. 44), and the recognition that legitimate and proper aspects of popular risk perceptions are people's perceptions of their own ability to access information and exert some 'control' over hazardous technologies. These are not external factors which *shape* perceptions of risk – they are elements *within* risk perception (Wynne, 1989a, 1989b). Thus 'technical' risk assessment, and concomitant notions of rationality, far from being an inexact science is actually fundamentally inadequate, since it fails to accord an a priori legitimacy to the voices of 'non-expert' groups and individuals. Treating critics – and potential critics – of the chemical industry as rational, and engaging in dialogue with them, means that the industry must transcend organizational exclusivity (Bellamy, 1984, p. 174; see also Edelstein, 1989) and existing forms of distorted communication (Habermas, 1970; Tombs, 1990). In other words, a greater input from groups currently defined as 'outsiders' into decisions surrounding the operation and control of hazardous technology is one key means both of achieving greater environmental protection and of altering popular perceptions of the chemicals industries.

These demands for empowerment – if taken seriously – require the development of both formal and informal mechanisms, at all levels, through which dialogue can be facilitated and fostered. Yet in considering empowerment, and the possibility of genuine forms of dialogue, the issue of information access clearly becomes crucial. While workers and publics may possess experiential knowledge, other information to which they would require access in order to enter into dialogue with the chemicals industries continues to be denied to them. These industries have traditionally guarded such data and, as we have indicated here, and as evidenced by the industries' opposition to the inclusion of strict right-to-know legislation in the new Environmental Protection Act, there are few signs of their shifting from this position. To the extent that they continue to do so, then any claims to be taking environmental issues and fears seriously remain hollow. Ironically, in the context of much of the industry's own rhetoric concerning the irrationality of many 'non-expert' groups, access to information on the part of such groups would in principle overcome the problem of 'ignorance' ascribed to them.

References

Bellamy, L. (1984) Not waving but drowning: problems of communication in the design of safe systems, in *Ergonomics Problems in Process Operations*, Institution of Chemical Engineers Symposium Series no. 90, IChemE, Rugby, pp. 167–77.

Bishop, D. R. (1985) The community's rights and industry's response, in M. A. Smith (ed.) op. cit., pp. 163–76.

Blowers, A. (1984) *Something in the Air. Corporate Power and the Environment*, Harper & Row, London.

Bowman, E. and Kunreuther, H. (1988) Post-Bhopal behaviour at a chemical company, *Journal of General Management*, Vol. 25, no. 4, July, pp. 387–402.

Brown, J. (ed.) (1989) *Environmental Threats: perception, analysis and management*, Belhaven/ESRC, London.

Bruel, J-M. (1990) Borrowing the land of our children, *Chemistry & Industry*, 19 November, pp. 734–6.

Chemical Industries Association (1986) *Chemicals in the Community*, CIA, London.

Chemical Industries Association (1989) *Total Quality Management*, CIA, London.

Chemical Industries Association/British Standards Institute (1987) *BS 5750: Part 2: 1987. the Standard and Guidelines for use by Chemical and Allied Industries*, CIA/BSI, London.

Chemistry & Industry (1989) HFAs get the industry's OK, 16 October, p. 659.

Chemistry & Industry (1990a) DuPont unveils 134a Plant, 1 October, p. 586.

Chemistry & Industry (1990b) ICI to end 'double standards', 17 December, p. 819.

Chenier, P. J. (1986) *Survey of Industrial Chemistry*, John Wiley, New York.

Clausen III and Mattson (1978) *Principles of Industrial Chemistry*, John Wiley, New York.

Cowe, R. (1991) Chemical formula for the environment, *The Guardian*, 12 April, p. 12.

Coxall, B. and Robins, R. (1989) *Contemporary British Politics*, Macmillan, London.

Crenson, M. (1971) *The Un-Politics of Air Pollution: a study of non-decisionmaking in cities*, Johns Hopkins Press, Baltimore.

Dewhurst, P. (1991) Letter to *Chemistry & Industry*, 18 March, p. 196.

Di Meana, C. R. (1990) Accepting the green challenge, *Chemistry & Industry*, 19 November, pp. 744–6.

Dispatches (1991) Tapping into Toxnet, Channel 4, 18 April.
Ecotec (1989) *Industry Costs of Pollution Control. Final Report to the Department of Environment*, Ecotec Research and Publishing Ltd., Birmingham.
Edelstein, M. R. (1989) Forcing a critical perspective on technology: the role of community opposition to facility siting. Paper presented at the Second International Conference on Industrial and Organizational Crisis Management, New York University, 3–4 November.
Foucault, M. (1980) Power and strategies, in C. Gordon (ed.) *Power/Knowledge. Selected Interviews and Other Writings 1972–7 by Michel Foucault*, The Harvester Press, Brighton. pp. 134–45.
Francis, M. (1990) Quality – the key to improved productivity, *Chemistry & Industry*, 1 January, pp. 13–15.
Gee, D. (1987) Letter to *The Guardian*, 24 October.
Grant, W., Paterson, W. and Whitston, C. (1988) *Government and the Chemical Industry: a comparative study of Britain and West Germany*, Clarendon, Oxford.
Habermas, J. (1970) Towards a theory of communicative competence, in H. Dreitzel (ed.) *Recent Sociology*, Macmillan, New York, pp. 115–48.
Henry, A. (1988) Communication concerning technological risk between industry and its surroundings at Pont-de-Claix, *Industry and Environment*, Vol. 22, no. 2, April/May/June, pp. 11–17.
Horton, R. (1990) Well begun is half done, *Chemistry & Industry*, 19 November, pp. 747–9.
Ilgen, T.L. (1983) Better living through chemistry: the chemical industry in the world economy, *International Organization*, Vol. 36, pp. 647–80.
Irwin, A., Smith, D. and Jupp, A. (1988) Audiences, disseminators and the social negotiation of uncertainty: understanding local technical issues. Paper presented at the 6th Management of Risk in Engineering Conference, London, 26–28 October.
Ives, J. (ed.) (1985) *The Export of Hazard*, Routledge & Kegan Paul, London.
Leonard, H. J. (1988) *Pollution and the Struggle for World Product: multinational corporations, environment, and international comparative advantage*, Cambridge University Press.
Liardet, G. (1991) Public opinion and the chemical industry, *Chemistry & Industry*, 18 February, pp. 118–23.
Lindheim, J. (1989) Restoring the image of the chemical industry, *Chemistry & Industry*, 7 August, pp. 491–4.
Malpas, R. (1987) The chemical industry and the environment, *Chemistry & Industry*, 21 September, pp. 643–6.
Marx, K. (1976) *Capital, Volume 1*, Pelican, Harmondsworth.
McGrew, A. (1992) The political dynamics of the 'new' environmentalism, this volume pp. 12–26.
Morrow, P. (1989) Developing responsible care, *Chemistry & Industry*, 1 May, pp. 279–80.
Pearce, F. (1990) 'Responsible corporations' and regulatory agencies, *The Political Quarterly*, Vol. 61, no. 4, pp. 415–30.
Pearce, F. and Tombs, S. (1989) Bhopal: Union Carbide and the hubris of the capitalist technocracy, *Social Justice*, Vol. 16. no. 2, pp. 116–45.
Pearce, F. and Tombs, S. (1990) Ideology, hegemony and empiricism: compliance theories of regulation, *British Journal of Criminology*, Vol. 30, no. 4, pp. 423–43.
Pearce, F. and Tombs, S. (1991a) Policing corporate 'skid rows': safety, compliance, and hegemony, *British Journal of Criminology*, Vol. 31, no. 4, pp. 415–26.
Pearce, F. and Tombs, S. (1991b) US capital versus the Third World: Union Carbide and Bhopal, in F. Pearce and M. Woodiwiss (eds.) *Global Connections: national and international aspects of crime and crime control*, Macmillan, London.
Perrow, C. (1984) *Normal Accidents*, Basic Books, New York.

Pettigrew, A. (1985) *The Awakening Giant. Continuity and Change in ICI*, Basil Blackwell, Oxford.
Popoff, F. (1989) Three tests for chemistry, *Chemistry & Industry*, 20 November, pp. 753–6.
Roddom, J. R. (1991) Letter to *Chemistry & Industry*, 18 March, p. 196.
Rose, J. (1990) Some have openness thrust upon them, *Chemistry & Industry*, 18 June, p. 376.
Sheldon, T. A. and Smith, D. (1991) Assessing the health effects of waste disposal sites: issues in risk analysis and some Bayesian conclusions, in M. Clark, D. Smith and A. Blowers (eds.) *Waste Location: spatial aspects of waste management, hazards and disposal*, Routledge, London.
Smith, D. (1990) The international trade in hazardous waste: a study in the geo-politics of risk. Paper presented at the ECPR Workshop on Crisis Management, Bochum, 2–7 April.
Smith, M. A. (ed.) (1985) *The Chemical Industry after Bhopal*, IBC Technical Services Ltd., London.
Stover, W. (1985) A field day for the legislators: Bhopal, and its effects on the enactment of new laws in the United States, in M. A. Smith (ed.) op. cit., pp. 69–126.
Todd, Lord (1987) Chemicals and the environment, *Chemistry & Industry*, 21 September, pp. 641–2.
Tombs, S. (1990) A case study in distorted communication, in *Piper Alpha – Lessons for Life-Cycle Safety Management*. Institution of Chemical Engineers Symposium Series no. 122, IChemE, Rugby, pp. 99–111.
Tombs, S. (1991) Injury and ill-health in the chemical industry: decentring the accident-prone victim, *Industrial Crisis Quarterly*, Vol. 5, pp. 59–75.
Trowbridge, M. E. (1987) Getting the balance right, *Chemistry & Industry*, 21 September, pp. 647–50.
United Nations Environment Programme (1988) APELL (Awareness and Preparedness for Emergencies at Local Level), *Industry and Environment*, Vol. 11, no. 2, April/May/June, pp. 3–7.
Viscusi, K. and Magat, W., with J. Huber, C. O'Connor, J. Bettman, J. Payne, and R. Staelin (1987) *Living With Risk. Consumer and Worker Responses to Hazard Information*, Harvard University Press.
Witcoff, H. A. and Reuben, B. G. (1980) *Industrial Organic Chemicals in Perspective, Part One: Raw Materials and Manufacture*, John Wiley, New York.
Woolard, E. S. (1990) A sustainable world, *Chemistry & Industry*, 19 November, pp. 738–40.
Wynne, B. (1989a) Frameworks of rationality in risk management: towards the testing of naive sociology, in J. Brown (ed.) op. cit., pp. 33–47.
Wynne, B. (1989b) Building public concern into risk management, in J. Brown (ed.) op. cit., pp. 118–32.

11

THE GREENING OF EUROPEAN INDUSTRY: What role for biotechnology?

Kenneth Green and Edward Yoxen

Introduction

This chapter examines the possible role of biotechnology in certain industrial sectors in the European Community (EC) in the 1990s.[1] Our argument is that much depends on how commercial organizations respond to continual controversy over how biotechnology is utilized. The analysis has two aspects: first, we look at the future of three industrial sectors and the likely uptake of biotechnological innovation in products and processes; and second, we look at how particular conceptions of nature mobilize sentiment in support of or in opposition to biotechnology.

We argue that environmentalism will be a very important factor. Hitherto, environmentalism has been considered mostly as an obstacle to technical change. But there has been a remarkable shift in attitudes towards the environment over the past few years, illustrated by the Europe-wide success of Green parties in the 1989 elections to the European parliament.[2] Many different forms of environmentalist sentiment are now evident. What was once viewed as extremism is now increasingly accepted as an important social phenomenon, to which European industry must respond. It is not so much that consumers are becoming Greens, but that 'Greenism' is being interpreted by commerce and industry as a new form of consumerism.

Changes of attitude, consumer behaviour and ideology will interact with structural changes in the system of production throughout the 1990s. Fordism was established as the paradigm of production and of consumption in advanced capitalist countries after the Second World War. Its production-technology basis has been the use of general-purpose machinery by skilled metalworkers to make standard parts in large quantities to be assembled into cheap uniform products by unskilled workers. This has been eroded by the automated and decentralized manufacturing made possible by micro-electronics and based on smaller production units and a more flexible use of labour.

This form of production organization has been dubbed 'flexible specialization'.[3]

Reprinted with permission of Butterworth-Heinemann from *Futures*, (1990) pp. 475–95.

The flexibility has the simple economic effect of reducing the cost of making smaller batches of products, in effect lowering the scale economy threshold. Changes in the product can also be more easily achieved, facilitating rapid product innovation. Thus flexible specialization both is a response to and lays the basis for increased competition and the increased power of the distributor. It allows retailers to track consumer behaviour more closely and to force manufacturers to supply goods targeted on particular market segments.

These trends are not at work with the same intensity in all industries. There are many where flexibility in production is severely limited, due to engineering problems. This is the case, for example, with the basic process industries – such as bulk chemical or basic foodstuff manufacture – which provide a range of inputs into other industries and for which rapid product change is neither necessary nor economically desirable, given the huge investment required to achieve the economies of scale. Nevertheless, many biotechnology innovations are particularly suited to the flexible specialization form of production and industrial organization.

Technical Trends in Biotechnology

The term 'biotechnology' as used here denotes a broad and heterogeneous field of applied science and related strategic research, that has arisen within the past twenty years. Its constituent sub-fields depend typically for their commercial appeal on a mixture of old and new expertise – such as controlled fermentation and novel sensor technology.[4]

Figure 11.1 shows which products form the bulk of biotechnology sales, in terms of both volume and value. Three main groups of products can be differentiated: *very high value* medical products, like vitamins and antibiotic drugs; an *intermediate* group of amino and organic acids, which are typically used in the animal feed and food industry or as chemical feed-stocks; and a group of *low value* products that have to be sold in enormous quantities and which compete against very similar commodities produced by different means.

Newer kinds of drugs, like the cephalosporins, are extremely expensive, but command a premium price because of their clinical advantages. They form a reference group for novel biologicals – like interferon, tissue plasminogen activator (TPA) and erythropoietin – that can be produced in recombinant micro-organisms. They are generally extremely difficult to isolate and impossible to synthesize because of their size as molecules. Their production in commercially viable quantities has only become possible in the past fifteen years with genetic engineering. Despite these technical advances the costs of production are high. One could also add to this product group the monoclonal antibodies produced by specially created cell lines, called hybridomas, which can be used in minute quantities in diagnostic kits or novel drug delivery systems or to image areas of tissue, but which also have a high unit cost.

A frequently expressed view is that these technologies – the genetic engineering of bacteria and yeasts and the culture of hybridomas to produce monoclonal

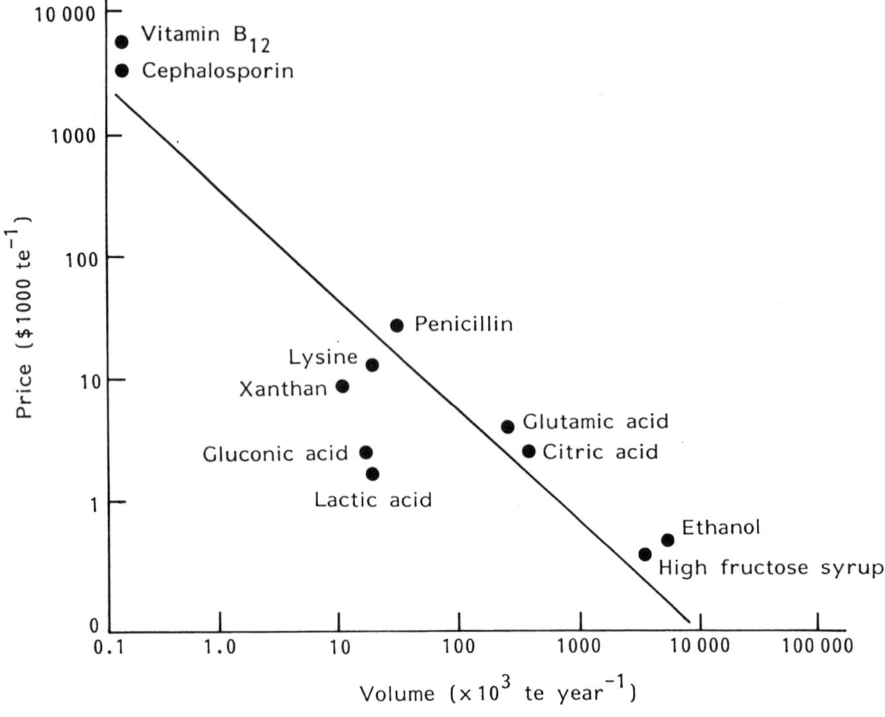

Figure 11.1 Biotechnology products – unit costs and production volumes.

antibodies – are the essence of biotechnology. However, as Figure 11.1 shows, such a view ignores two major product areas, and much else besides.

Thus at the other end of the scale – in price and volume terms – there lie fermentation processes, such as that for producing ethanol as a bulk industrial chemical. Similarly, the production of corn syrups now occurs on a large scale, despite the production quotas set in certain areas like the EC. These novel products have to compete with traditional ones which are often highly protected, so process innovations are vital.

A range of organic chemicals comes in between, among them amino acids such as glutamic acid or lysine. Again, biotechnological production has to compete against other processes. The novel foods like Rank Hovis McDougall's fungal protein, Quorn, used in savoury pies, belong here too, as do the bacterial cultures being developed as soil inoculants, to protect plants from pests or to enhance their growth by supplying additional nitrogen to the roots.

Despite its enormous promise, biotechnology is a field beset with a number of major technical and economic problems, such that investment in some areas can be extremely risky. The massive investment by ICI in single-cell protein production failed because of the rise in energy costs and the falling prices of competing soya products (Hacking, 1986, pp. 98–102). There are also enduring *regulatory* uncertainties, even with products which have been on the market for some years but are produced by traditional means.[5] In addition, the whole area of *intellectual property rights* is one of enduring controversy, not just over the finer details of patenting procedures and their international comparability but over major issues of patentability *per se*.[6]

Indeed, contemporary biotechnology is irredeemably controversial. To understand this we need to review the impact of biotechnology on particular industries in Europe – chemicals, food and agriculture, and health care. We describe the structure and place of each within the European economy and the controversies either that biotechnology might engender or, indeed, for which it might provide a solution.

Chemical Industry

World production of chemicals in 1987 was worth over US$800 billion ($10^9$). Western Europe had the largest share of that production (30 per cent), with North America second (25 per cent). Japan and the newly industrializing countries (NICs) have rapidly increased their share since 1974 (from 20 per cent to 28 per cent). The NICs' production is likely to continue to grow fastest over the next ten years so that by 2000 it will approach that of Western Europe, although the mix of chemicals will differ greatly (Metz, 1988).

Western Europe is also a major exporter – 15 per cent of its production is sold outside its region; the equivalent figures for the USA and Japan are 8 per cent and 9 per cent respectively. Although in most engineering-based industries Western European firms play second fiddle to US and/or Japanese transnationals (TNCs), this is not the case with the chemical industry. Nine of the world's fifteen largest chemical firms are based in EC countries, including the four largest – Germany's BASF, Hoechst and Bayer, and the UK's ICI.

All the large companies, whatever their national origin, are active competitors in the European market. Over the past few years, the EC firms have in turn become more active in the USA, buying up US chemical companies, small and large, for their technologies and their production and marketing facilities. European firms are also becoming more involved in Japan. In turn, the Japanese are getting involved in production and marketing in Europe: Mitsubishi Gas Chemical markets a polymer in a joint venture with Solvay whilst Sekisui Chemical and Kureha Chemical both have manufacturing facilities in the Netherlands.[7]

The industry has two main divisions – 'traditional' production of organic and inorganic basic 'building-block' chemicals in bulk; and knowledge-intensive speciality chemical products, like pharmaceuticals, pesticides, detergents, dyestuffs, photochemicals, specialist fibres and plastic materials. By the late 1970s, there was substantial world overcapacity in basic chemicals; with the slump in the early 1980s, the industry in Europe was forced to close plants and lay off large numbers of workers. The big European firms made strategic decisions to reduce their dependency on basic chemical production. They took the view that much of world demand would be met by the expanding basic chemical industry in the Middle East and in the East Asian NICs, and decided to switch their emphasis towards global niche markets for research-intensive speciality chemicals and to working with customers to develop higher-value applications of under-exploited existing products.

The chemical industry has always been a high spender on R&D, at a level two to three times higher than manufacturing industry in general. In 1987 Bayer spent over $1.1 billion on R&D, or 6 per cent of sales. Hoechst spent the same; BASF spent $750 million, 4 per cent of sales; ICI spent $700 million, 4 per cent of sales.[8] As production has shifted to new product areas, so has R&D.

This switch to value-added chemicals exemplifies the move towards 'flexible specialization'. Basic chemical mass production, which will remain a major part of European chemical production, is inevitably associated with inflexible capital-intensive techniques. Speciality chemicals, on the other hand, are produced in smaller quantities, in smaller plants; they can thus be more quickly modified as user demands change. Conventional process innovation is still important, not least in energy saving and in the reduction of pollution.

The European chemical industry expects to benefit greatly from the advent of the single European market and is well prepared for it.[9] Its large firms are placed to take rapid advantage of the elimination of non-tariff barriers. However, non-EC chemical TNCs are also preparing themselves. US firms are looking round for small and medium-sized hi-tech European firms with which to establish joint ventures. Japanese firms have begun to base low-labour-cost manufacturing facilities in Spain and Portugal to take advantage now of subsidies which are less likely to be on offer in the future when the EC adopts uniform but less favourable regional policies. We should not assume that a Japanese offensive in the chemicals industry will be an unqualified success; the European chemical industry is stronger and better equipped than the European automobile/electrical engineering industry was.

Biotechnology has much to offer the chemical industry. In processes, enzymes have considerable potential as biological catalysts, although they are restricted to low-temperature fermentation processes. Bulk production of basic chemicals uses non-biological technologies that are so efficient that it is highly unlikely that biotechnologies could ever replace them. However, treatment of chemical wastes by microbial means is likely to prove a useful application of biotechnology R&D. Micro-organisms exist that are able to degrade a wide range of highly toxic chemicals, which are often difficult to store and dangerous to burn. A number of specialist companies offer contract research and disposal services, in what is called bioremediation. Major chemical companies are themselves developing new processes for waste treatment. In the Netherlands Gist-Brocades claims to have established a leading position with its bacteriological effluent treatment process, with twenty plants in operation world-wide.[10]

Criticism of the chemical industry over pollution issues has been mounting over the past few years, fed by incidents such as the spillage of chemicals into the Rhine by the Swiss company Sandoz. The industry is thus under pressure to reduce its emissions of polluting chemicals into water and the air, and to find new ways of dealing with hazardous solid waste. Harmonized EC-wide emission standards are being negotiated and, given the strength of the German chemical industry, the standards are likely to be similar to those already in place in Germany. Since these are already the toughest in Europe – Germany's spending on environmental protection almost doubled from 1982 to 1987 – the chemical industry in other EC countries will be put under greater competitive pressure. Some relocation may result. Yet the industry still seems resigned to stricter controls over its activities. Investment in pollution control equipment, averaging 15 per cent of total chemical industry investment in 1987, is likely to rise (Marsh, 1989).

Environmentalist pressure on the chemical industry to alter its products is increasing. The most notable recent example has been the campaign against chlorofluorocarbons (CFCs). The manufacturers of CFCs (ICI, DuPont and Allied Signal/Atochem) are engaged in intensive research into ozone-friendly alternatives. Other environmental 'hit-list' chemicals include persistent pesticides, nitrate fertilizer and a variety of food additives (*Economist*, 1989). The industry's shift to high-value, flexible specialization has made it easier to envisage incorporating environmental considerations in some of its new products.

Biotechnology thus offers the chemical industry considerable innovative scope. However, environmentalist/green objections have already been raised to some biotechnology initiatives. Hoechst has been involved for about two years in a legal battle with a German local action group of radical environmentalists who are objecting to the construction of a factory to produce genetically engineered human insulin. Recently, the Administrative Supreme Court in Hesse held that there is no legal basis on which the safety of such work can be assessed and halted production plans.[11]

It is of course quite possible that the more radical forms of green opposition to biotechnology will not endure. Instead, a generalized less militant 'greening' of the European population, expressed in its continued demands for a cleaner environment, could stimulate and legitimate some applications of novel biotechnology.

However, another scenario is that environmental action groups might remain at loggerheads with the chemical industry and might gain support among the European population as it becomes more environmentally aware. On this view, relatively minor product changes will be insufficient to calm environmentalist fears and biotechnology will be seen as contributing to continuing environmental despoliation. In this case, the rates of diffusion of new biotechnologies would be significantly slowed. Since Germany is the European country with the most politically entrenched green movement (electorally represented at European, federal, provincial and local levels), it would be Germany where a slowing in innovation would be most likely. Given the importance of the German chemical industry this would have a serious effect on the competitiveness of the European chemical industry as a whole. As a consequence, the chemical industry could well shift the focus of its activities to countries where regulations are weaker and public criticism less forceful, possibly in Eastern Europe. However, even here environmental sentiment is being expressed much more strongly and environmentalists are frequently closely involved with political reform. At the same time these countries are desperate to attract foreign investment. This suggests that one should think carefully about how environmentalist and technocratic sentiments operate in different cultures and how persuasive different conceptions of future industry are likely to be.

To recapitulate, it is useful to consider the future of the chemical industry with a simple two-dimensional graph. First, the strategic pressures are shifting the industry away from mass production of basic organic and inorganic chemicals to flexible production of higher-value 'designer' chemicals. This is the vertical dimension of Figure 11.2. Numerically this could be the percentage of turnover obtained from mass or flexible production. Second, the growing environmental objections to the various aspects of chemical production could shift companies away from traditional industrialism in favour of some kind of green approach. Thus the horizontal dimension of Figure 11.2 indicates the degree to which chemical companies have accommodated themselves to strong environmentalist concern. This is harder to put into numbers.

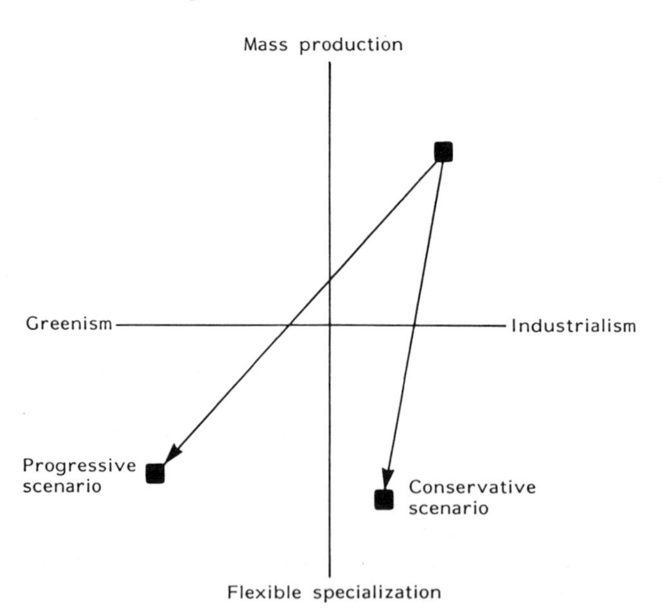

Figure 11.2 Scenarios for the chemical industry.

We can place the chemical industry of the post-war period in the top right quadrant at present. 'Industrialism' has strong economic and social roots in European countries and it is perfectly feasible to envisage greenist trends being only a minor irritant to the industry as it moves away from concentration on bulk chemical production. Thus one scenario (a 'conservative' one) would see the chemical industry of the early twenty-first century certainly diversifying into novel areas with higher-value products and possibly making substantial improvements in process innovation and pollution control, without real accommodation to greenism. One might see the relocation of largely unmodified hazardous production to more permissive EC countries, or to Eastern Europe, with continuing commitment to highly chemical-intensive agriculture. On our grid the conservative scenario shifts the locus of the industry to the bottom right quadrant.

Another possibility is that flexible specialization becomes more important and that major changes in production take place which represent major concessions to greenism. This is the bottom left quadrant. The other credible possibility is no movement at all, which could be the case if present experimentation were transient.

Agriculture and Food Industries

The industrial processing of raw food materials is a major economic activity in EC countries; in 1985, over 2 million workers produced over $240 billion worth of processed food in EC12. This amounts to about 14 per cent of the EC's manufacturing output, although in some EC countries the share is much higher (Stevens, 1987). The agricultural and food processing industries are obviously closely connected – the output of the former being the raw material for the latter. Yet in their structure they are considerably different: in all EC countries, food processing as a manufacturing industry is a major employer but highly oligopolistic. The same is hardly true of agriculture – in many EC countries, there are still large numbers of small producers of basic foodstuffs, even if farms are getting larger, fewer in number, and employing ever fewer farm workers.

Agriculture in Europe has long been the site of political controversy focused on the twin requirements of guaranteeing Europe's food supply and ensuring adequate remuneration for the politically important agricultural population. Food processing, whether in terms of the structure of the industry, the increasingly capital-intensive techniques it employs, or its science-intensive products, has for the most part been free from such controversy. However, it seems likely that 'food' as a whole, its quality and means of production from the farm to the table, will become more controversial over the next ten years. The applications of biotechnology to both agriculture and food processing therefore will be subject to a more intense public gaze than would have been expected ten years ago.

Agriculture

The future development of European agricultural production is strongly dependent on the set of subsidies, regulations and institutions that make up the EC's Common Agricultural Policy (CAP). Its effectiveness is much disputed (Duchene, Szczepanik and Legg, 1985). Some see it as nothing more than an agency for the protection of inefficient, if numerous and therefore politically important, farmers; in their view the CAP has raised European prices of food and industrially important raw materials over world market levels. Others see the CAP as having been too strongly in favour of large-scale capitalist farming with its attendant overemphasis on energy-intensive and (environmentally damaging) chemical-intensive farming techniques.

Whatever the contributory role of the CAP, no one can doubt the success of European agriculture in providing more than enough food for Europe since the 1950s – a success perversely celebrated in the food surpluses that dominated public debate about the CAP in the 1980s. There has been a rethink of European agricultural policies so as to reduce surpluses, switch subsidies to the poorer regions and introduce alternative products.[12] As for the latter, the prospects for switching production from one agricultural item in surplus (such as wheat or milk) to another not in surplus (e.g. sunflower seeds) are limited, since consumption of the alternatives is not commensurate. There is thus an opportunity for considering new products altogether (*Financial Times*, 1988, 1989a). Candidates for this are peas and broad beans as protein sources for animal feeds, reafforestation (to cope with the EC's likely shortage of wood in the next century), renewable raw materials for industrial use or for energy generation (biomass, possibly using new plants particularly suitable for this purpose) and, most significantly for our argument, organic crops for the green consumer. Farmers who are best able to respond to these more rapid demand changes will be those whose farming methods are more adaptable – this could be seen as 'flexible specialization' applied to agriculture.

Overproduction, brought about by the CAP, and the related need to curtail steadily increasing costs of the policy as a whole, formed the background for an interesting instance of a failed biotechnological innovation in the agricultural area, namely bovine somatotrophin (BST). This hormone controls the conversion of animal food into milk in cows. Its administration by injection or implantation increases production of milk for a given quantity of food.

By the early 1980s a range of major chemical companies, including Monsanto, were gearing up for a major sales drive of this new product, by organizing tests on farm animals, to which BST was administered on a trial basis to obtain regulatory approval. The UK was selected as the test country in which marketing approval was to be first sought, leading, it was hoped, to approval throughout the EC.

BST gradually began to encounter more and more criticism. The criticisms focused on three issues. First, there were those who claimed that sufficient amounts could be ingested by drinking milk to cause harm to babies and children. Second, some claimed that cattle treated in this way were in fact subject to veterinary problems and the use of the product was therefore inhumane. Third, it was clearly hard to deny that BST would exacerbate an already severe problem of overproduction of milk, which in 1986 led governments across the EC to implement cuts in milk production quotas and to take other measures, equally problematic politically, to encourage farmers to produce less milk. One could argue that in the longer term any measure that allowed producers to be more productive would be advantageous to them, even in a saturated market, but would still require farmers to reduce their herd size and to farm with more scientific control over their activities. Many small farmers felt that this was beyond their means.

In 1988 the UK regulatory authorities decided not to approve the marketing of BST at that point. They took the view that the scientific evidence showed that cows were adversely affected by use of the hormone and that this was not acceptable. By mid-1989, the evaluation of BST at the European level was still in process, with the EC's Agricultural Commissioner indicating that a decision might take a further eighteen months to emerge.

A wide range of groups have concerned themselves with the issue, for many different reasons. The outcome was unexpected to many, and certainly went against the declared expectation of the UK agriculture minister. It also showed real differences of perception within the farming community in the UK. The debate about BST was at one level a struggle between competing ideas about 'natural' limits of agricultural animals and the value of different farming practices. The introduction of BST failed because its proponents were unable to

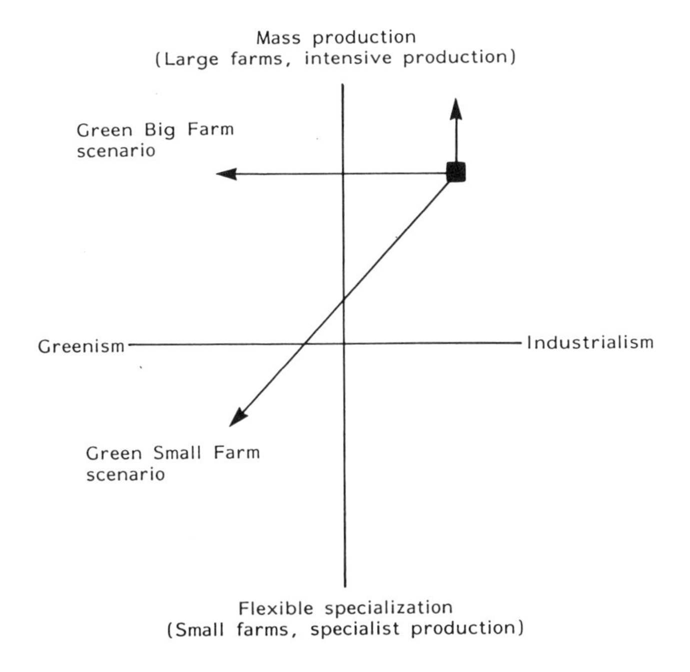

Figure 11.3 Scenarios for agriculture.

build sufficient support for this elaboration of intensive dairy farming.[13]

Similar considerations may well apply to a technology already under development as a possible sideline for farmers, the production of high-value biologicals, like clotting factor, from transgenic animals (*Sunday Telegraph*, 1988).

The same scenario framework that was used in our discussion of the chemical industry can be applied again. As before, in Figure 11.3 the vertical dimension represents structural change. Historically, the movement has been from small-scale production to larger and larger units, to reach the upper right quadrant.

One possibility is continued concentration, without accommodation to environmentalist pressures. This is indicated by the vertical arrow in the Figure, moving upwards. Another possibility is no change in overall scale of production, but with much greater environmental awareness – represented by a switch to biological pesticides, greater conservation efforts, less nitrogen input, and increased on-farm digestion of organic wastes. This is the horizontal arrow in the Figure, and it is a very likely possibility. We call this the 'Green Big Farm' scenario.

Another scenario can be envisaged, however, with much greater diversification of production, combined with a change of scale as small farms find it possible to occupy emergent market niches, with a wide range of new products grown without massive chemical inputs. This scenario might be one that many radical green activists envisage, with undertones of a traditional, even feudal, past, where farming was based on intuitive environmental understanding and experience rather than driven by advanced science. We call this the 'Green Small Farm' scenario. Much will depend on the structure of subsidy and encouragement to diversify and on the growth of new markets for agricultural commodities. For example, the future of organic farming probably lies with the purchasing decisions of supermarket chains.

Food processing

Science-based innovations in food processing have produced the huge range of food combinations on sale today. Processed foods have increasingly expanded their share of total agricultural/foods production and trade. About one-third of total agricultural/food trade between OECD countries is now in processed foods; intra-EC trade in processed food has increased tenfold since the 1960s and the EC is now a net exporter of processed food.

Although there are many small and medium-sized firms in the industry, particularly producing 'speciality' foods, as a whole the industry is fairly concentrated. R&D expenditure in the food processing industry is not very high in any country. For OECD countries it averages 0.8 per cent of output, compared to 4 per cent for manufacturing as a whole. R&D is concentrated on product innovation, as product lifecycles are especially short in what is a highly competitive industry. Interestingly, Japan's commitment to food-related R&D (measured per capita of population) has been more than twice that of the EC (Stevens, 1987).

Biotechnology in the EC food processing industry raises three issues:

1. costs of raw materials
2. rate of product innovation
3. greater consumer awareness of quality

These are discussed in turn below.

Costs of raw materials

The outputs of agriculture account for between two-thirds and three-quarters of the value of food processing sales. Any cost-reducing new technologies have enormous appeal. The food industry wants to ensure that appropriate new crops and methods of plant and animal husbandry diffuse as fast as possible into European agriculture. Inevitably this means that it has a keen interest in the workings of the CAP, an interest which is not always the same as that of the farmers.

The battle over high-fructose corn syrups illustrates this well. These have diffused rapidly into the US food and drinks industry. In the EC, on the other hand, the beet growing and processing lobby has been able to ensure that strict production limits exist.

Rate of product innovation

Product competition and increasing pressure from retailers are pressing food firms to speed up their rate of product innovation. This fits well with trends in new technologies of food production, particularly in final product formulation and in packaging. The mass production continuous flow sectors of cereals processing, sugar refining and brewing have been the site of major automation developments from the 1950s onwards. So far, some of the most economically significant applications of biotechnology have been in these sectors, permitting the use of different, cheaper raw materials. More recently, the introduction of the enzyme chymosin, produced in genetically engineered bacteria, has speeded up the maturation of Cheddar cheese. Similarly, in brewing, using new strains of yeast has allowed both more economical use of established raw materials and energy saving.

However, in those sectors closer to the retailing end, and thus more directly subject to changing consumer choices, particular foods have to be produced in smaller batches; so more labour is required here, as it is in packing and inspection. Developments in flexible automation, robotics and computerized weighing and inspection are thus of crucial importance to the economics of the industry. However, demand for basic processed foods is nowadays inelastic with regard to price, so competition takes place over shifts in consumer tastes and patterns of purchasing. The ageing of the population, the increased participation of women in waged work and the smaller size of household all change the kinds of food we are prepared to consume. The widespread use of kitchen equipment such as freezers and microwave ovens obviously affects the types of food purchased and presents firms with opportunities to present new added-value food reformulations. Much of the resulting product innovation is of a minor kind. But firms can also incorporate recent R&D results more rapidly. Changing consumer views of what kinds of food are healthy and objections to added chemicals can also be a source of innovation as food companies modify their products to emphasize how 'natural' they are – the new active yoghurts are an example – or to remove sugar, salt and cholesterol, or add fibre.

Greater consumer awareness of quality

Increased consumer knowledge is often a product of campaigns for stricter regulation of all types of food additives. Such campaigns are likely to intensify over the next few years. Biotechnology offers solutions to some of these problems. Various new polymers are being promoted as fat substitutes. Dipeptide sweeteners, like aspartame, are being advertised as health-promoting. Rapid tests, based on monoclonal antibody (MAb) technology, to identify the presence of dangerous infections in foodstuffs will soon be in commercial use. Recent scares, particularly in the UK, concerning *salmonella* infections in eggs and chickens, and *listeria* infections in cook-chill foods and soft cheeses, have shown up the inadequacies of current assays, using selective culturing and agar plating, which can take up to thirteen days to give a result. MAb-based tests can reduce this to thirty hours (Green, 1989b).

The single European market will be of major significance for the future of the food industry, and the strategic moves to profit from this arrangement will have major consequences for food production, distribution and retailing. Thus a major part of the harmonization of regulations in preparation for a single market has been concerned with food regulations, and the removal of trade barriers based on traditional product definitions. The classic instance is the definition of beer in Germany in the Reinheitsgebot. In general, regulatory comparability has been achieved by removing or weakening the prevailing national composition requirements, with major consequences for food standards (*Financial Times*, 1989a).

At the same time companies have sought to prepare themselves for the intensified competition post-1992, and for Community-wide operation through acquisitions and bilateral agreements. The French company BSN, the second biggest EC-owned food manufacturer, has acquired the UK sauce company HP/Lea and Perrins, the Belgian brewer Maes, the Italian brewer Peroni, and the Italian/Spanish grocery firm Star. Three Danish food firms recently merged, to become one of the top five EC-owned food companies, and the Spanish sugar company Ebro has acquired other Spanish and Portuguese food manufacturing and distribution firms. The Japanese conglomerate Mitsubishi recently acquired the British canning company, Princes Foods, the first of an expected wave of East Asian food manufacturers moving into Europe (*Financial Times*, 1989b).

Interest in food quality and composition, as an adjunct to the environmental movement

This is the area where notions of the 'natural' are most obviously at work, as food manufacturers seek to understand and manipulate consumers' shifting ideas about food that is intrinsically health-promoting. This set of trends links both agriculture and the food sector. One possibility is that agricultural diversification will be accelerated by supermarkets deliberately promoting organic foods, grown by accredited suppliers without the use of chemical fertilizers and crop protection products. Another is that economic pressures on farmers will force more and more of them to experiment with novel plant and animal species, such as llamas for wool production, or flax for novel industrial materials. Clearly, genetic engineering has a role to play here.

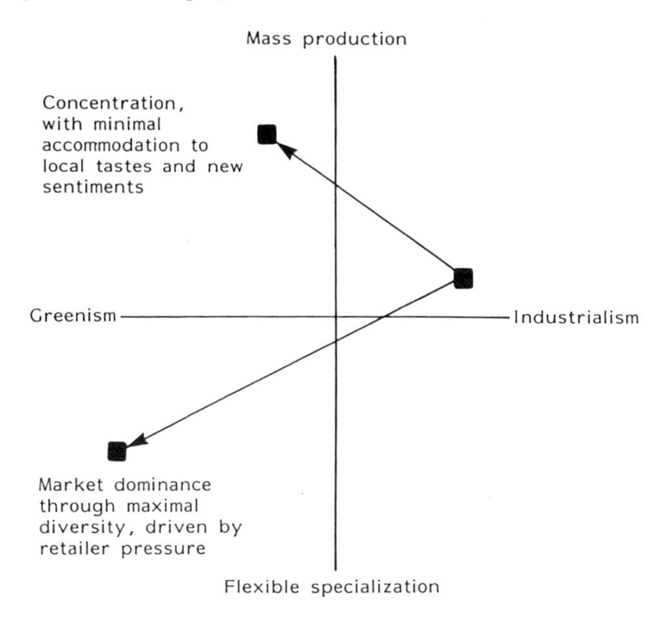

Figure 11.4 Scenarios for the food industry.

Yet again, the two-dimensional schema serves to indicate possible paths of development (Figure 11.4). Food manufacture in Europe is becoming more concentrated as the pressures to achieve market dominance intensify. At the same time there is no doubt that food manufacturers are taking seriously consumer interest in environmentalism and 'natural' food products, at least in the sense of developing additional product lines that capitalize on consumerist notions of the 'natural' and 'environmental-friendliness'.

Quite how the cultural phenomenon of green consumerism unfolds – whether it will lead many people on to more radical environmentalism, as Green parties' activists hope, or whether it will allow the incorporation and neutralization of radical environmentalism – remains to be seen. But whatever outcome is realized, some form of biotechnology will play an important facilitating role.

Health Care and Pharmaceuticals

'Health care' covers a large number of human activities. 'Formal' health care is what is provided by the organized health services – the primary health care clinics, hospitals and other organizations for care and cure and the agencies of preventive medicine. Such services can be delivered by public or private organizations, funded by private or state insurance schemes; they are all firmly regulated by governments through official health services and ministries of health.

'Informal' health care comprises three main parts – 'alternative' medical practitioners, the self-medication products sold in pharmacies and those products for self-diagnosis, and, of most significance, unpaid care of the sick and infirm. The distinction between formal and informal health care activities is especially important when we are considering the future development of health care in Europe. Because of the increase in the number of aged in the European population, informal health care is likely to grow in importance. But the boundaries between formal and informal care shift over time and innovations in biotechnology in health care are implicated in the shift.

The proportion of GDP spent on formal health care within the EC thus ranges between 4.6 per cent (Greece) and 9.1 per cent (France), the total amounting to over $250 billion in 1984. About 5 per cent of the EC's waged workforce is employed in the health services. About half of health service expenditure is on the hospital sector and about 10 per cent is on prescribed pharmaceuticals (OECD, 1987). Sales of proprietary medicines are about one-sixth of prescribed pharmaceuticals, making the European market worth ECU6.5 billion/year (Commission of the European Communities, 1989).

The future organization of health care will depend on cost containment, an increased emphasis on prevention and self-help, the impact of population ageing and changes in disease patterns, and the introduction of new health care technologies, not least biotechnology-based ones.

Since the mid-1970s the focus of governments has been on preventing costs rising faster than GDP. This has changed the way health services are organized. Health care financing agencies have looked harder at new ways of delivering health care, in particular more home nursing and the replacement of inpatient by outpatient services. In addition, governments have emphasized personal responsibility for illness prevention.

Population ageing in the OECD countries has yet to have a major impact on health care expenditure, although the numbers of the older aged are increasing rapidly. Shortages of nursing and paramedical staff are likely to be more of a problem over the next ten years. Closely connected with the ageing phenomenon is the changing pattern of disease among the European population. The rising number of aged also brings an increase in non-fatal deteriorations of the musculo-skeletal and nervous systems – from arthritis to Alzheimer's disease.

World sales of health care biologicals run at about $150 billion/year, with the lion's share of that going on pharmaceuticals. Consumption of prescribed (ethical) drugs is rising fast in the richer countries, with markets expanding particularly rapidly in the USA and Japan[14]. The pharmaceutical industry is strongly international with the largest firms active in health care markets world-wide. It costs over $100 million to research, develop, test and innovate a new drug. As a result, the pharmaceutical industry is highly research-intensive, with the major companies typically spending 10–15 per cent of their sales on R&D.

Biotechnology is particularly applicable to health care products including pharmaceuticals, vaccines and diagnostic kits. It also provides ways of more rapidly screening potential pharmaceuticals, speeding up and cheapening the expensive process of pharmaceutical innovation.

For diagnosis, a large number of new methods of testing human fluids and infections have been developed, based on MAb technology. The fastest growing diagnostics markets are in immunology – tests based on antibody–antigen reactions detecting hormones and markers for cancers and infectious diseases – and microbiology, testing for the presence of bacteria, viruses, parasites and fungi. MAb diagnostics are now used routinely in hospital laboratories worldwide, and there is a growing market for them in some European countries in primary health care clinics as well. MAb-based tests sold in pharmacies for confirming pregnancy are an established do-it-yourself market, and other over-the-counter (OTC) products have been introduced for monitoring fertility. MAb products are also a vital part of the growth of new types of imaging. In the longer term, the diagnostics companies promise accurate, rapid and cheap tests based on DNA probe and biosensor technologies which will detect genetic abnormalities and cancer markers. For therapy a whole range of new biotechnology-based drug delivery systems are under research, such as directed anti-cancer drugs which use MAbs to identify and kill cancer cells.

Biotechnology is poised to make a considerable impact on health care in the EC, while cost containment policies also force major changes. There are broader social movements that are critical of the curative bias and bureaucratic organization of the formal health services. This creates opportunity for the health equipment and pharmaceutical industry to offer products that assist self-diagnosis and self-treatment.

Also, any increased commitment of health service agencies to extensive screening would offer markets for new diagnostic and therapeutic products. Thus a major state-sponsored screening programme of the over-40s, for example, for heart disease might expand markets for anti-cholesterol drugs. This type of 'prevention' strategy would thus be more lucrative than life-style changes as far as the pharmaceutical industry is concerned. However, proposals for such screening programmes are already proving controversial in professional circles.[15]

Diagnostic products are only lightly regulated, since they are not consumed internally. However, their use could well come under closer scrutiny for broader public health reasons. The UK government has already announced that it will not permit quick tests for HIV (the 'AIDS virus'), based on MAb technology to be sold to unlicensed users. This outlaws the sale of do-it-yourself testing kits for HIV in pharmacies. The UK government is taking the same view on such tests as the US Food and Drug Administration (FDA), which has already refused to license tests which could be sold directly to physicians to perform the tests in their own office laboratories, a large potential market in the USA. Other European countries are apparently considering following the UK lead (Green, 1989a).

More immediate, however, are the continuing campaigns by green political groups against certain biotechnology-related health care products and medical procedures. We have already mentioned the campaign against Hoechst's plans for a human insulin manufacturing facility. There are also more widely based campaigns, with the support of many religious groupings, against new reproductive and 'human engineering' technologies such as contraceptives, abortifacients, *in vitro* fertilization and gene therapies.[16] Here, too, ideas about nature play an important role in the formulation of positions and the mobilization of sentiment.

The impact of the single European market on health care is difficult to predict. In the short term, the only obvious impact of 1992 would be to make it easier for the manufacturers of non-prescribed, OTC drugs to expand their markets. Health services themselves will continue to be 'unharmonized', and the process of harmonization of regulation of prescribed pharmaceuticals has been, and is likely to continue to be, slow. Also slow will be the realization of a uniform European pricing policy for pharmaceuticals – though if a drug were to cost the same in every EC country this would reduce the pharmaceutical industry's profits and might speed further concentration and restructuring of the industry.

To recapitulate: it is tempting to use once again the schematic two-dimensional diagram with which different scenarios have been depicted. In the case of health, however, this is not appropriate, in that we are analysing service institutions rather than supply industries. Even though we have concerned ourselves with structural change in the health care supply industries, the primary concern has been with the delivery of care through the hospital system and in the community. Given the diversity of institutions and systems it makes little sense to generalize about overall trends concerning these institutions.

Instead, it is important to reiterate the systematic effects of cost containment, as throughout the world health care costs continue to rise, driven by the increase in technological sophistication, by low productivity growth in the delivery of services and by population ageing. Biotechnologies will find a wide range of applications, in diagnosis, therapy and preventive medicine, in the near future. Although it is generally assumed that the major applications lie in hospital medicine, there are also important uses in alternate site diagnosis (e.g. in GPs' surgeries) and in home health care. Many of these applications pose moral dilemmas. Here they are highlighted only in the case of the diagnosis of HIV seropositive status, but gene therapy would provide another example. Both of these represent attempts to push the latest techniques into immediate application, although there may be a strong case for not doing so. Judgements on these questions will be handled differently in different cultures.

The Broader Picture

The establishment of a single market in the EC by the end of 1992 will have a major impact on the European economy and on its industrial and business structure (Cecchini, 1988). Of course, whether EC-country-owned firms make the most of the opportunities offered by market integration depends very much on the strategies which they pursue. Part of that response will involve thinking globally, but acting locally, using manufacturing systems that permit flexible specialization.

The exploitation of biotechnology has now to be seen in the context of larger European markets than would have been expected – the rewards to companies for successful innovation are thus much greater and the incentive for inter-firm collaboration in R&D is high. There can be no doubt as to the intensification of competition between TNCs within Europe – be they European-, US- or Japanese-owned. In many industrial sectors, the presence of Japanese firms, through either imports or, more recently, direct production in European subsidiaries, has forced changes in the structure of European firms, whether locally owned or subsidiaries of established US TNCs. So far, however, European-US-Japanese competition has been restricted to the automobile, office systems and consumer electronics sectors. In food and chemicals, competition in Europe has been solely between European and US firms. Although this is likely to continue, Japanese firms might soon become active in Europe in these sectors as well. The options open to European firms in these sectors will depend on how agriculture evolves.

Two issues will dominate agricultural development in the EC in the 1990s, the CAP and agricultural diversification. Attempts to reform the CAP will continue, because of its enormous cost, the grotesque anomalies of vast food surpluses, and continuing skirmishes with other major exporters of agricultural commodities. Similarly, the CAP countries will struggle to hang on to traditional European markets for sugar, cotton, oilseeds and rice.

Social attitudes to farming and farmers are changing. An increasing number of people see that, in some EC countries, many farmers receive substantial aid from national and Community funds, that some have become very wealthy as a result, that the heavy use of chemicals has environmental and long-term medical consequences, and that other ways of farming are perfectly feasible. It is not just from the exchequer that pressure to change farming will arise. Farmers themselves differ over commercial strategy and environmental management, which weakens them politically.

Biotechnology connects with these developments in a number of ways. It could exacerbate the problem of overproduction – through increased milk yields from supercharged cows, better varieties of crop species, more intensive horticultural production requiring less land. It can also suggest some solutions – such as novel high-value crops providing new industrial materials, production of biologicals from genetically engineered farm animals, new waste treatment systems that generate energy and reduce discharge of effluent into water courses. The problem is a mismatch in rates of change, as the need for agricultural reform and diversification runs ahead of the new technical possibilities.

In the 1980s, Green parties have become politically significant in a number of EC countries. The 1989 European elections showed that the Green parties and their allies now have electoral support in all twelve countries of the EC. But whatever the formal parliamentary *representation* of Green parties, the environmental policy issues that are their mainstay have struck a chord with European citizens of all political colours. As we have indicated in our sector studies, over the past few years – the Chernobyl disaster can be seen as a sort of watershed here – a number of environmental issues have thrust themselves on to the European political agenda: the pollution of the Rhine, acid rain, concern over damage to the ozone layer and over the disposal of hazardous wastes are but four.

As we have seen, industry is responding positively to the greening of the European population in ways large and small. The application of biotechnology will obviously be of some use in responding to green objections to the environmental unfriendliness of the current products and industrial processes. For example, biotechnology will provide new methods of waste treatment to reduce noxious emissions into water and the air. But equally, there are some biotechnology applications which will themselves be the object of green complaints. What is certain is that the industrial application of new biotechnologies will take place in the context of much more public criticism and scepticism than was the case with the introduction of new chemical technologies in the 1950s and 1960s.

The past fifteen years have seen much frustrating, frequently unproductive positional debate, on such matters as microbiological laboratory safety, the patenting of publicly funded research, the contribution of biotechnology to the continuing economic dependency of less developed countries, and the risks to the environment posed by the deliberate release of genetically engineered organisms. Underlying the public debate over various forms of biotechnology since the mid-1970s have been radically different ideas about living nature and the extent and form of acceptable interaction with it.

Some conception of nature is essential for any scientific investigation to take place, although different areas of science typically rest on different notions about nature. As important, many political ideologies incorporate explicit references to 'natural' emotions, or human nature, or the natural state of human beings that must be constrained by social rules. Ideas about nature are tremendously potent elements in political rhetoric of all kinds to maximize their appeal to particular constituencies.[17]

But even in science-oriented societies these underlying notions often come into conflict with those making up particular areas of scientific inquiry or technological innovation. Biotechnology is a particularly striking example. It calls into question the many different classifications of nature and underlying assumptions about aspects of the natural world. These include the notion of species as distinct entities, the distinction between natural objects and the ownable products of invention, and the idea that natural ecosystems are well understood or at least can be perturbed minimally in an experimental fashion to explore the effects of releasing new species into them.

What is so remarkable about such controversies is how people can be motivated by what look like somewhat abstract critiques of a certain kind of technology. Producing human insulin would seem to be a beneficial act, intended to earn a profit to be sure, but arguably preferable to building a nerve gas plant for a foreign power, or selling dangerous pesticides to countries with minimal regulations. Yet in the controversy over Hoechst's human insulin plant, a whole approach to medical treatment and the complex set of relationships between hospital medicine and capital-intensive research has been effectively called into question through the mobilization of sentiment that is avowedly critical of modern industry.

This is one area where subtly different conceptions of nature are put to work to mobilize sentiment, both for and against industrial innovation. The radical green critique, often with religious or metaphysical underpinnings, generally presents nature as a fragile system, constantly vulnerable to chemical agents that escape control (Rudig, 1985; Hay and Haward, 1988). Other more consumer-oriented rhetorics, readily adaptable to the language of marketing, represent nature as a state that can be emulated or recreated by scientific product development. The question is whether the frequent repetition of this latter theme in advertising and political rhetoric reinforces the more radical forms of environmentalism or undercuts them. Much may depend on the roots of such sentiment and campaigning, in particular national cultures and traditions.

There have been few analyses of the underlying presumptions, agendas, cultural resonances and rhetorical gambits of this now familiar, but endlessly frustrating, mode of political debate. Why do issues like those just mentioned engender such powerful and complex feelings? Do the debates evolve? What do the participants learn as the years of largely inconclusive argument go by? What is its deeper cultural significance? In other words, how can meaningful and productive participation in industrial, commercial and governmental decision-making concerning biotechnology come about? There is evidence that 'greenism' is increasingly recognized by industry as a potent new form of consumerism. Some feel it needs to be read correctly and used as a source of competitive advantage, by enabling successful appeals to be made to more discerning consumers. We believe that the anticipation of consumer/activist sentiment is becoming more important in a changed economic climate, and that without it moves towards more flexible production systems, which biotechnology could facilitate, will be constantly checked.

Notes

1. The chapter is adapted from Yoxen and Green (1989).
2. Green parties, running under that name or allied with other radical leftist or regionalist parties, got over 8 per cent of the votes cast in Europe as a whole (fourth behind the Socialists, Christian Democrats and Communists). Strictly as 'Greens', they ranged from 15 per cent in the UK to less than 1 per cent in Greece.
3. 'Flexible specialization' is discussed in Piore and Sabel (1984) and in Blackburn, Coombs and Green (1985); for a review of 'flexspec' arguments see Allen and Massey (1988), pp. 136–83; for an analysis of 'flexspec' in Europe see Lane (1988) pp. 141–68.
4. For discussions of the technical and industrial aspects of biotechnology, see Jacobsson, Jamison and Rothman (1986); Hodgson (1989); Marx (1988); and Yoxen (1986).
5. Regulatory problems posed by biotechnology are summarized in OECD (1989) pp. 56–62.
6. For an account of the main issues at stake in patent protection in biotechnology, see ibid., pp. 63–4.

7. *Chemical and Engineering News*, 26 September 1988, pp. 11–12; other European-Japanese joint ventures are listed in van Tulde and Junne (1988) pp. 240–2.
8. *Financial Times*, 7 September 1988.
9. As Gunter Metz of Hoechst, president of CEFIC, the European chemical industry federation, puts it, 'Most of the major chemical companies anticipated European integration as early as the 1960s – despite the barriers to trade at that time. Europe is our home market. The direct implications for us are not as numerous as in other industries, where deregulation will almost amount to a revolution' (Metz, 1988).
10. Reuters Textline, news item, 26 September 1988.
11. *Chemistry and Industry*, 4 January 1988, p. 7.
12. For accounts of the reform of the CAP, see Harris, Swinbank and Wilkinson (1983) and Tracy (1989).
13. There is as yet no full account of the BST controversy; our account is derived from news reports and journal discussions compiled from an online database search; see news reports in *The Independent*, 14 September 1987, p. 15; *New Scientist*, 1 September 1988, p. 3; and 'Europe delays BST decision', *Nature*, Vol. 340, 1989, p. 415; A product that's 'too effective', *Chemical Week*, Vol. 138, 1986, pp. 10–11.
14. Thus from 1976 to 1985, per capita consumption of pharmaceuticals in the USA jumped from \$36 to \$111, in Japan from \$36 to \$116; for EC countries the rises were smaller, from \$55 to \$98 in Germany and from \$51 to \$81 in France; see *Financial Times*, 8 November 1988, Supplement p. 4.
15. For a review of the arguments for and against screening for cholesterol see the papers in King's Fund Forum (1989).
16. Gene therapy is a procedure for correcting human genetic diseases by getting functional human genes into human somatic cells; the first officially sanctioned experiment in human gene therapy in the USA started in 1989; it involves introducing a genetic marker (antibody resistance gene) into activated lymphocytes which are used in new forms of cancer therapy. See *Business Week*, Weatherall (1988).
17. For discussions of the uses made of 'nature', see Douglas (1970); Midgley (1978); Jordanova (1986).

References

Allen, J. and Massey, D. (eds.) (1988) *The Economy in Question*, Sage, London.

Blackburn, P., Coombs, R. and Green, K. (1985) *Technology, Economic Growth and the Labour Process*, Macmillan, London.

Business Week (1989) Human gene therapy: after a lot of looking, now the leap, 1 May, p. 133.

Cecchini, P. (1988) *The European Challenge 1992: The Benefits of a Single Market*, Wildwood House, Aldershot.

Commission of the European Communities (1989) *Panorama of EC Industry 1989*, Office for Official Publications of the EC, Luxembourg.

Douglas, M. (1970) *Natural Symbols*, Barrie and Rockcliff, London.

Duchene, F., Szczepanik, E. and Legg, W. (1985) *New Limits on European Agriculture: Politics and the Common Agricultural Policy*, Croom Helm, London.

Economist (1989) A biodegradable recyclable future, 7 January.

Financial Times (1988) Italian palate may soon be sampling West German pasta, 15 July, p. 3.

Financial Times (1989a) West Germany: court rules against ban on foreign sausages, 3 February, p. 2.

Financial Times (1989b) 18 April, supplement on the food industry.

Green, K. (1989a) HIV-OTC test ban, *Biotechnology Insight*, no. 68, January, pp. 4–5.

Green, K. (1989b) Listeria hysteria: biotechnology to the rescue? *Biotechnology Insight*, no. 69, February, pp. 3–4.

Hacking, A. J. (1986) *Economic Aspects of Biotechnology*, Cambridge University Press.
Harris, S., Swinbank, A. and Wilkinson, G. (1983) *The Food and Farm Policies of the European Community*, John Wiley, Chichester.
Hay, P. R. and Haward, M. G. (1988) Comparative Green politics: beyond the European context, *Political Studies*, Vol. 36, no. 3, pp. 433–48.
Hodgson, J. (1989) *Biotechnology: Changing the Way Nature Works*, Equinox, Oxford.
Jacobsson, S., Jamison, A. and Rothman, H. (1986) *The Biotechnological Challenge*, Cambridge University Press.
Jordanova, L. J. (ed.) (1986) *Languages of Nature: Critical Essays on Science and Literature*, Free Association Books, London.
King's Fund Forum (1989) Blood cholesterol measurement in the prevention of coronary heart disease. Sixth Consensus Development Conference, King's Fund, London.
Lane, C. (1988) Industrial change in Europe: the pursuit of flexible specialization in Britain and West Germany, *Work, Employment and Society*, Vol. 2, no. 2, pp. 141–68.
Marsh, P. (1989) The chemical industry: in a very tight corner, *Financial Times*, 21 April, Industry and Environment Supplement.
Marx, J. L. (ed.) (1988) *A Revolution in Biotechnology*, Cambridge University Press.
Metz, G. (1988) 1992 – the CEFIC viewpoint, *Chemistry & Industry*, 5 December, pp. 744–8.
Midgley, M. (1978) *The Beast in Man*, Cornell University Press, Ithaca, NY.
OECD (1987) *Financing and Delivering Health Care*, OECD, Paris.
OECD (1989) *Biotechnology: Economic and Wider Impacts,* OECD, Paris.
Piore, M. J. and Sabel, C. F. (1984) *The Second Industrial Divide*, Basic Books, New York.
Rudig, W. (1985) The Greens in Europe: ecological parties and the European elections of 1984, *Parliamentary Affairs*, Vol. 38, no. 1, pp. 56–72.
Stevens, C. (1987) Technology and the food processing industry, *STI Review*, no. 2, September, pp. 7–40.
Sunday Telegraph (1988) UK Frankenstein farming condemmed, 18 September, p. 3.
Tracey, M. (1989) *Government and Agriculture in Western Europe 1880–1988*, Harvester Wheatsheaf, Hemel Hempstead.
van Tulde, R. and Junne, G. (1988) *European Multinationals in Core Technologies*, Wiley/IRM, Geneva.
Weatherall, D. (1988) The slow road to gene therapy, *Nature*, Vol. 331, pp. 13–14.
Yoxen, E. (1986) *The Gene Business: Who Should Control Biotechnology?* Free Association Books, London.
Yoxen, E. and Green, K. (1989) Scenarios for biotechnology in Europe: a research agenda. Report submitted to the European Foundation for the Improvement of Living and Working Conditions, December.

12

THE FRANKENSTEIN SYNDROME: Corporate Responsibility and the Environment

Denis Smith

Introduction

> How can I describe my emotions at this catastrophe, or how delineate the wretch whom with such infinite pains and care I had endeavoured to form?
>
> *(Mary Shelley (1818)* Frankenstein, *p. 105*)

As previous chapters in this book testify, environmental issues are currently riding high on political agendas within Europe and North America. This concern follows in the wake of chronic pollution episodes (such as acid rain, global warming and the destruction of the ozone layer) and also from a series of major accidents involving either the production, storage or transport of chemicals. These incidents, which include Seveso, Bhopal, Piper Alpha and the *Exxon Valdez*, illustrate the need for society to be able to exert some form of control over the activities of industry. They also confirm the necessity to make corporations accountable to those who may have to bear the costs of any impacts that arise from industrial production.

Public pressure over environmental issues has created a new set of constraints acting upon industrial production. Decision-makers can no longer assume that the public will tolerate continuing environmental exploitation and degradation. This book has sought to address some of the fundamental issues facing business within the context of this 'new environmentalism'. Central to our discussion has been the role of corporate responsibility, operating within the context of a range of environmental concerns, in providing the framework within which such issues can be assessed. This final chapter seeks to draw together some of the main strands from the various functional areas of business that have been outlined earlier and, as a result, aims to explore more fully the process of corporate responsibility.

Corporate responsibility is not a new concept. Its roots can be found in the eighteenth-century concerns over the impact of industry on the environment and in terms of worker health and safety. Society has always expressed its concerns over those aspects of its activities that it cannot fully control. During the period of the industrial revolution, writers often used images of evil and despair to describe the activities of industry. Anyone reading Blake's description of the 'dark satanic mills' can be left in no doubt as to the negative image that was being attributed to the workings of industrial enterprise. Writers also used biblical analogies, during this period, to describe the environment within northern working-class towns like Widnes and Runcorn. One such writer lamented the demise of local environmental standards by observing that,

> it is permissible to imagine that the ashes of Sodom and Gomorrah immediately after the destruction of those wicked cities emitted no more pestiferous odours, or were more devoid over all their wide area of one single green leaf or sign of vegetation, than those twin towns of the Mersey – Widnes and Runcorn.
>
> *(Fisher Unwin, 1888)*

Perhaps the most poignant image of our failure to control science and technology can be found in Shelley's Frankenstein. In 1818, Shelley was asking the fundamental question that many environmentalists are still seeking to answer today concerning the long-term impacts of technological developments. This question centres around the impacts of our scientific and technological activities and focuses our attention on the long-term damage that may result from short-term gain. For the current issues of waste disposal and global environmental degradation, such inter-generational effects are of prime importance. In many respects the current wave of environmental concerns are reminiscent of this public fear of 'science' and technology. Certain sections of industry, notably nuclear power, chemicals production and waste disposal, have become the 'Frankenstein' of the twentieth century; indeed, the names Bhopal and Chernobyl have taken over the stigma associated with the Frankenstein mantle. Even today, the main concern that is being expressed by many public and environmental groups still relates to our ability to effectively control that which we have created to serve us and upon which we have become so dependent.

As a consequence of the 'doomwatch' mentality that exists amongst certain sections of society, even the more 'benign' areas of human and industrial activity, such as the disposal of domestic waste products and the problems associated with low-level pollution, have become important political issues which have to be addressed by both government and industry. Conflicts surrounding the purity of drinking water, product contamination and the incineration of domestic waste, have all attracted considerable media attention. These incidents, acting in combination with the more acute and catastrophic events listed earlier, have raised questions concerning society's ability to control the actions of the corporation. This issue of the control of corporate activities lies at the heart of the current debate concerning business and the environment. In order to be able to frame the issues raised in earlier chapters of this book we have to explore in more detail the meaning of the term 'corporate responsibility'.

Corporate Responsibility – Towards a Paradigm Shift?

Modern society presents business with immensely complicated problems. Technology

> has advanced to a level that tests intellectual capabilities, markets have become more complex and international in scope, and difficult new problems of social issues and social responsibility have arisen.
>
> (*Davis, 1975, p. 19*)

The impacts of industry have persisted as a dominant theme within the social sciences throughout the twentieth century. Central to society's concern over environmental degradation have been the notions of corporate control and responsibility, with the latter growing in importance during the 1980s. During this period there has been considerable debate over the precise meaning of the term (see De George, 1986) and Sethi (1975, p. 58) has gone so far as to observe that it has

> been used in so many different contexts that it has lost all meaning. Devoid of an internal structure and content, it has come to mean all things to all people.

Despite such concerns it is possible to highlight the main boundaries of the concept: and, whilst there is debate over the extent of corporate obligations, a number of writers have provided the framework in which the main issues can be addressed.

The perceived obligations of business within a framework of corporate responsibility have been widely debated within the academic literature and cover a range of activities from worker welfare through product liability to environmental impacts (Carroll, 1979). However, the sheer range and complexity of the issues, raised by the question of corporate responsibility, combine to make the precise definition of the term difficult. In addition, there are a number of competing 'political' perspectives set out by proponents and opponents of the corporate responsibility process (see Walters, 1977). These divides can occur both within and between particular political perspectives and serve further to muddy the waters of definition. Much of this dispute centres around industry's reluctance, and that of its supporters, to accept a broad-based definition of its responsibilities. Consequently, industry has generally sought to challenge any move to impose such a 'burden' upon it.

With these problems in mind it is necessary to attempt to chart out the main boundaries of the term. Jones (1980, pp. 59–60), for example, argues that within a framework of corporate responsibility it is incumbent upon corporations

> to have an obligation to constituent groups in society other than stockholders and beyond that prescribed by law or union contract.

He develops this by arguing that the obligation has to be voluntary and the scope of the responsibility must be so broad as to go beyond the traditional stakeholder concepts. In developing this theme, Murray and Montanari (1986) define the nature of a socially responsible firm as one which is seen to accomplish 'the desired ends of society in terms of moral, economic, legal, ethical, and discretionary expectations' (p. 816). For some writers, however, such a definition would be too broad in that it goes beyond the economic and legal limits which were often held to mark the boundaries of the firm's responsibility. Notable amongst such objectors is Friedman (1962) who sees the responsibility of the firm purely in legal and economic terms.

Within the context of our current discussions, the process of corporate responsibility can be considered at three distinct levels (Sethi, 1975). At its most basic we expect industry to conform to its legal and economic obligations to society. Within this phase corporate legitimacy is determined as a result of market forces and legal interpretation, along the lines advocated by Friedman (Sethi, 1975). However, as earlier chapters have indicated, a reliance on market forces to control pollution has failed in the past and the legal system is only as effective as the methods used to enforce the law (see Chapter 4). Obviously such an essentially reactive strategy does not allow for the lag between social requirements and legislative change and is therefore a process which will attract considerable conflict around corporate activities.

Given the concern over the boundaries of responsibility we need to ask the question, how far should we expect industry to go within the context of its behaviour? For some writers the answer is, 'as far as it takes'. Davis (1975), for example, argues that the obligation of the business decision-maker is to improve the 'quality of life' enjoyed by society, in addition to maximizing his/her own business interests. The question of incompatible interests is obviously raised at this juncture and it is this which bedevils the whole process of industry–public interactions. The difficulty here lies in the constantly changing nature of societal norms for, as Sethi argues,

> A specific action is more or less socially responsible only within the framework of time, environment and the nature of the parties involved. The same business activity may be considered socially responsible at one time, under one set of circumstances and in one culture, and socially irresponsible at another time, in another place and under different circumstances.
>
> *(Sethi, 1975, p. 59)*

Because of the problems of predicting the nature of future societal demands, many corporations have been reactive to such pressures rather than seeking to be proactive. Indeed, there are numerous cases in which corporations have sought to undermine the basis of societal demands by using technical expertise, or some other such power base, to deny the existence of an environmental problem (see Crenson, 1974; Blowers, 1984; Smith 1990a, 1990b). Whilst there are obvious exceptions to this restricted approach from some companies, such as the Body Shop, many of the improvements that have been made in environmental policy have arisen from some form of 'external' pressure, usually articulated in the form of industry–public conflicts.

Such basic behavioural requirements from industry are patently not enough to arrest the degree of damage that is currently being inflicted on the environment by corporate and other activities. As a result, society demands more than mere conformity with the law, or a reliance on the market, and this moves us into Sethi's second level, namely corporate (social) responsibility. Here the firm seeks to go beyond that prescribed by law and contract and instead seeks to conform to the current demands of society. Here Sethi defines the process of corporate responsibility as

> bringing corporate behaviour up to a level where it is congruent with the prevailing social norms, values, and expectations of performance

and thus

> While the concept of social obligation is proscriptive in nature, the concept of social responsibility is prescriptive in nature.
>
> (*Sethi, 1975, p. 62*)

This process of being, as Sethi puts it, one 'step ahead of time' lies at the centre of the current debates over the environment, as industry seeks to ensure that it remains *au fait* with the short- to medium-term demands of the public. The importance of such a strategy lies in Sethi's assessment that a failure of industry to meet societal demands has been the root of most of the industrial conflicts during the post-war period.

However, a major problem can exist if corporations are merely seeking to prevent more rigorous regulatory intervention by simply 'appearing' (through increased public relations or the utilization of technical expertise) to be socially responsible. Unfortunately, the 'burden of proof' in such cases often lies with those who might be affected by the actions of the corporation. Power, expressed in economic, legal or technical terms, can often be brought to bear on a particular conflict to support the corporate view of the legitimacy of their actions against the expressed concerns of affected publics. However, such a short-sighted approach by industry is unlikely to serve its interests in the long term as it would become increasingly embroiled in conflicts surrounding its activities. Indeed, industry has begun to recognize that a responsible approach can serve its interests better than a process which simply ignores the wishes of society. What is obviously required here is a much more strategic view of the problem and this lies at the heart of Sethi's final stage, namely social responsiveness.

Within this phase, industry takes a proactive role regarding environmental quality and attempts to prevent problems from arising rather than just seeking to respond to them. For Sethi the issue here is

> not how corporations should respond to social pressures, but what should be their long-run role in a dynamic social system.
>
> (*Sethi, 1975, pp. 62–3*)

In other words, industry should take account of the likely societal demands when formulating their strategies and re-order their priorities as a consequence. However, the reality of the situation is that many organizations are still too reactive and seek to 'handle disturbances' rather than think strategically about such problems (Mintzberg, 1989). This situation has certainly prevailed within the area of environmental concerns as, until recently, few organizations took 'green issues' into account when formulating their strategies. As Shrivastava observes in Chapter 3, the dominant definition of the environment has been configured purely in economic terms with scant attention being given to the concept of a 'green competitive advantage'. Such a narrow view of the corporate environment has been widely criticized by writers who espouse a more value-driven view of corporate strategy (see Freeman and Gilbert, 1988; Freeman,

Gilbert and Hartman, 1988; Gilbert, 1986). The basis of this argument lies in the belief that, because values are inherently important in determining the choices that managers make when formulating strategies, then strategy itself is inherently bound into a whole series of ethical issues. Whilst some organizations are presently attempting to think strategically about environmental issues, many more are unfortunately simply conforming with current societal norms, or worse still, appearing to do so in an attempt to gain a competitive advantage in the short term.

Drawing on the three levels outlined by Sethi, it is possible to illustrate the complex nature of corporate 'responsibility'. The various stages of business 'obligations' to society are shown in Figure 12.1, where it can be seen that there

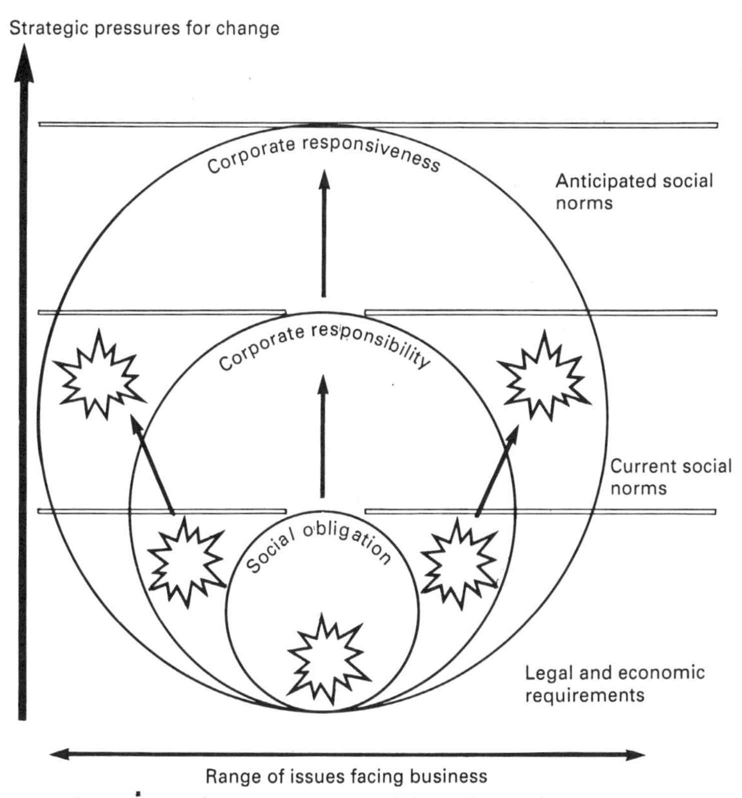

Figure 12.1 The concept of corporate responsibility, after Sethi (1975).

is the potential for conflict at each of the three levels. In terms of the corporation's economic and legal obligations conflict can occur over the interpretation of such obligations. It is possible for the more powerful groups in society to challenge the interpretation of the law and such challenges have been mounted by industry in the area of environmental issues. For example, certain organizations have used their superior resources to challenge those environmental demands, rather than to conform to the wishes of local publics and thereby act in a socially responsible manner. At the level of corporate responsibility, such challenges are easier to mount by those groups who are opposed to the activities of the corporation, because of the greater number of issues over which the corporation is held to have some responsibility. Due to the changing nature of societal norms the corporation has to be extremely flexible in its approach to environmental issues and can no longer be seen to 'hide' behind current regulations. The potential for conflict is also present within Sethi's last level, that of corporate responsiveness. Here the corporation has to predict the likely demands that will be placed upon it over a longer time span. As a consequence, the potential for underestimating such requirements can lead to the incubation of crises which will emerge at a later time. This strategic dynamic of the problem is an important consideration which needs to be developed further.

Despite disagreements concerning the boundaries of the responsibility, a common strand in most definitions is the notion of obligation and managerial action, whether this is expressed in the narrow terms of shareholders, as argued by Friedman, or in the wider sense as used by authors such as Sethi (1975), Ackerman and Bauer (1976), Carroll (1979) and Murray and Montanari (1986). It is in terms of the sheer range of activities over which the term spans that has prompted some writers to go so far as to claim that the issue of corporate responsibility is subversive (Friedman, 1962).

Irrespective of the motives for corporate responsibility it is apparent that it spans a number of areas of industrial activity. Carroll (1979) offers a framework for the analysis of corporate performance across a range of activities. The model has three main components. The first relates to the responsibilities of the organization in terms of its economic, legal, ethical and discretionary responsibilities. The second component concerns the social issues involved within the process and these include the environment, product and occupational safety, in addition to consumerism, discrimination and shareholder interests. The final element in the matrix relates to the components of organizational response, reaction, defence, accommodation and proaction which Carroll takes from the work of Wilson (1974). Within this context Carroll argues that

> The assumption is made here that business does have an essential responsibility and that the prime focus is not on management accepting a moral obligation but on the degree and kind of managerial action.
>
> *(Carroll, 1979)*

The resultant model is shown in Figure 12.2 and serves to provide a framework for an analysis of the issues raised during many of the previous chapters in this book.

Within the context of earlier discussions in this book it is apparent that the range of social issues facing business varies according to the sector. The chapters by Green and Yoxen, Tombs and Essery illustrate how such factors vary across sectors and also how different 'stakeholders' within a sector can perceive a problem. Fischer and Shrivastava point to the different strategies that have been adopted by business to cope with the new pressures that impact upon them. In the recent past the actions of a considerable number of corporations have been in terms of reacting to pressure and using technical expertise to defend against the demands made by stakeholder groups. Both Fischer and Shrivastava argue

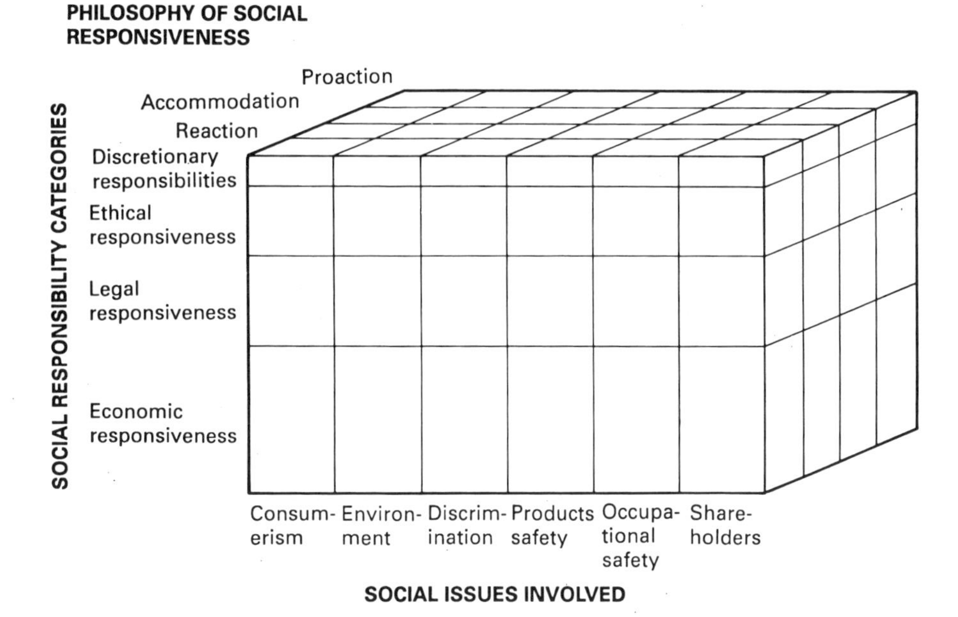

Figure 12.2 The corporate social performance model. Source: Carroll (1979).

the case for a more open and participatory form of expertise to allow corporations to move away from a narrow consideration of economic and legal responsibilities. However, such a quantum change in the base operating philosophy of many corporations will not come easily and we need to address the vexed question of how we ensure that the organization conforms to the demands placed upon it by the new environmentalism.

Corporate Responsibility in Strategic Perspective

What drives a firm towards the process of corporate responsibility? Evidence from the past decade would suggest that corporations have had sufficient warning concerning the negative impacts that a constrained view of their responsibility brings with it. There is no denying the importance of corporate responsibility within the present political climate. The only defence that many companies now offer relates to the need for a 'level playing field' so that all their competitors will be forced to comply with the same standards of behaviour. Within well-defined political groupings, such a process should be relatively easy to implement. However, the problem is more difficult to resolve at the international level and particularly between developed and developing countries (see Smith and Blowers,

1991). Whilst steps have been taken to deal with this issue, an effective regime for enforcement is required to ensure success.

The important point to be made here is that notions of ethical behaviour underpin the strategy process as decision-makers tend to act in accordance with their own value systems (Goodpaster, 1983). Freeman and Gilbert (1988) outline two fundamental principles within this approach – the values principle and the interdependence principle. The basis of these principles lies in the assumption that all actions taken by the organization and its constituent members are underpinned by the values held by those individuals and groups (values principle). Consequently, any actions taken by them will involve decisions or choices which, in turn, will be affected by the dominant value system (interdependence principle). Therefore, Freeman and Gilbert argue, the process of strategic management must take formal account of the values held by members of the organization and other stakeholders in influencing policy formulation and implementation. As such, the strategies adopted by corporations should rest upon sound ethical principles and decision-makers should attempt to make the organization both more democratic and less analytical in order to cope with the pressures of an ethically aware regime. To this end, Freeman and Gilbert (1988, p. 11) argue that

> Corporate strategy which ignores the role of people in the organization simply ignores why organizational members act in the way they do. Corporate strategy must return to the individual values of corporate members, before it is formulated. It must be built on these values rather than taking them as constraining forces.

Despite the persuasive power of such reasoning the academic study of strategic management is still dominated by the economically driven models of competition and seems reluctant to take on board many of the principles of an environmentally ethical approach (see Chapter 3).

The necessity to strive towards legitimacy is a powerful force acting on the corporation and requires careful management in order to ensure that the corporation does not get overtaken by shifts in public opinion. However, the inability to effectively predict such societal shifts is a process that has thwarted the ambitions of a number of companies over the last decade. Who could have predicted the rapid shift away from CFCs as propellants within aerosols? The threat to CFC producers is obviously apparent as their market begins to shrink rapidly. But such shifts in society's preferences also bring with them opportunities for development as the search for an ozone-friendly CFC substitute testifies. Whilst economic self-interest is obviously a prime motivator in this case we need to ascertain what other factors, if any, drive companies towards social responsibility.

Driving Forces for Corporate Responsibility

Mintzberg (1983a) argues that social responsibility is manifest in a number of ways: responsibility for its own sake, enlightened self-interest, sound investment theory and the circumvention of outside influence. There is little evidence to suggest that many firms have adopted a strategy of corporate responsibility for its own sake, particularly in the area of environmental impact (see Smith, 1991b). However, where it does occur, the basis of this approach entails corporations anticipating and responding to future environmental pressures arising from industrial activities and taking the necessary preventative measures to ensure compliance (Mintzberg, 1983a). Whilst such a socially responsive approach is indeed rare, there are cases of social responsibility which is driven by notions of self-interest (in its various guises) and this is typified by the remaining elements of Mintzberg's driving forces.

Enlightened self-interest, for example, assumes that benefits will accrue to the organization, largely in an indirect manner, as a result of behaving in a responsible manner. Within the context of the new environmentalism (see Chapter 2), the sheer range and magnitude of societal environmental concerns provide us with a framework within which enlightened self-interest takes on board a new momentum. One only has to look at the marketing strategies currently employed by a number of washing powder manufacturers to see the importance of wider social concerns in product promotion. It is now almost inconceivable for such a product to be marketed without reference to its environmental benefits, even though such benefits may be difficult, if not impossible, to prove. Hajer (1990) observes that this re-emergence of public environmental concern stems from a combination of realigned social values and a recognition of the fallibility of technocracy. Whilst the realignment of social values has created opportunities for industry, the critique of technocracy strikes at the heart of the corporate power base.

The roots of the challenge to technocracy can be found in a number of major accidents and pollution episodes that took place during the 1980s which have had serious repercussions for the industrial concerns who were responsible for them. The most serious of these was the accident in December 1984 at Bhopal which resulted in the immediate deaths of at least 2,500 people and has led to the delayed deaths of some 300 affected people per year since the release. The effects of the accident on Union Carbide's reputation were understandably great and resulted in widespread public opposition to its existing and proposed sites around the world. In addition to the Bhopal disaster, there were a series of other accidents during this period at Mexico City (1984), Chernobyl (1986) and Basle (1986) which, along with increasing concern over the transfer of hazardous waste to the developing world (1988) and the depletion of the ozone layer, raised the public consciousness of the potential impacts of industrial production. Within this context the power of 'enlightened self-interest' has been considerable. The Bhopal accident, for example, prompted the development of the 'community care' programmes in the USA and, to an extent, in the UK, and led decision-makers to recognize that the sector as a whole had to ensure that there was no repetition of the disaster. Whilst there is evidence of the operation of an enlightened approach by the chemicals industry in the wake of Bhopal, it is also possible to see evidence of Mintzberg's remaining categories, namely 'sound investment theory' and 'interference avoidance', often operating in a synergistic manner.

Sound investment theory is based on the premise that the value of an organization will be affected by the market's perception of its social behaviour (Mintzberg, 1983a, 1983b). The basis of this proposition is that the market will reward behaviour that is perceived to be in the interests of 'society', and conversely will punish behaviour that is seen to be detrimental (Mintzberg, 1983a). An example of the latter can be found in the suggested loss of £24 million off the share price of the waste disposal company ReChem, during a period of environmental conflicts surrounding attempts to import Canadian PCB wastes into the UK (see Smith, 1991b). Here, the market appears to have expressed concern over the legitimacy of the company and its activities set within the context of a hostile social environment. The combination of local public opposition, actions by the dockworkers and, perhaps most importantly, negative media coverage, served to damage the market's confidence in the company. Similarly, the case of the Union Carbide Corporation, in the wake of the Bhopal accident, illustrates the point that the corporation may also have to pay the costs of its activities directly. In this case Union Carbide were forced to divest a considerable amount of their assets in order to prevent a hostile take-over attempt (Shrivastava, 1992). Whilst Union Carbide paid compensation to the Indian government for the disaster, the sum involved was pitifully low by comparison with what would have been due had the accident occurred in a western country. In the long run, however, Union Carbide's position was strengthened after the event as the market saw its apparent ability to cope effectively with a crisis event of this magnitude. Today the company claims it is leaner and fitter than it was in the period leading up to the disaster. However, the company will for ever retain the stigma associated with the accident and it may have been in an even stronger position had the disaster not occurred.

The notion of advantagous effects arising from corporate behaviour is a key element within the process of the sound-investment theory. Here the basis of the argument is

> that one behaves responsibly not because of ethics – because that is the 'proper' way to behave – but because it is to one's advantage to do so.
>
> *(Mintzberg, 1983a, p. 4)*

In both the ReChem and Union Carbide cases it is apparent that different approaches to the problems besetting the companies would have resulted in more favourable economic and political outcomes.

The concept of advantage is also a key underpinning mechanism of Mintzberg's final argument, namely avoiding interference. Here the aim of the corporation is to avoid political interference, expressed in terms of pressure group activity or increased regulatory pressure from the government, although it is argued that in the process rhetoric, rather than action, often ensues (Mintzberg, 1983a). There is a fundamental flaw in each of the self-interest arguments, according to Mintzberg, as they illustrate the necessity for more rigorous external controls on the organization. What is required, therefore, is that we strive to change the culture of the organization so that it acts in a socially responsive manner as a matter of course. However, such a process is difficult to achieve in the short term; therefore we need to explore the control opportunities that are available to regulators.

How Do We Control the Corporation?

> the works ... were belching forth volumes of most deleterious gases and clouds of black smoke from chimneys of inadequate height, with trees that stood leafless in June, and hedgerows that were shrivelled in May. The air wreaked with gases offensive to the sight and smell, and large heaps of stinking refuse began to accumulate.
>
> *(Allen, 1907)*

The extent of environmental degradation has been a powerful driving force behind the move towards controlling corporate activities. Whilst we are no longer subject to environmental degradation on the scale described by Allen, we still have to deal with a range of problems which have the potential for widespread damage to the physical environment. The question still remains, however, as to how we can achieve this control in an effective manner. This issue has been addressed by a number of writers, but perhaps most notably by Mintzberg (1983a; 1983b; 1989) who sets out the control options for industry in the form of a conceptual horseshoe (see Figure 12.3). If we move through the horseshoe it is possible to assess the relevance of each of the measures for the control of environmental degradation.

The concept of *nationalizing* the firm would, on the face of it, offer little scope in allowing for environmental improvements. The Central Electricity Generating Board as was, and now Powergen and National Power, have shown scant regard for the environment of Scandinavia through their 'tall stacks' policy of pollution dispersion. Similarly, the state's control of industry in what was the Soviet Union has illustrated that pollution problems are not determined by state ownership and political ideology. In the UK, a further example of the failure of 'nationalization' in controlling pollution can be found in the nuclear industry, which has both been a major polluter of the Irish Sea and generated considerable concern over the possible link between nuclear power and leukaemia, especially amongst children. In all of these cases the national (economic) interest has taken precedence over the concerns of local publics and, in some cases, other national governments.

If nationalizing the organization cannot arrest the flow of pollution then perhaps a *democratizing* process can be more effective? Within this context Mintzberg argues for greater accountability and openness within the corporate decision-making process in an attempt to overcome the narrow definition of the organization's 'interest'. However, as Smith (1991b) observes, such a process is fraught with difficulties in terms of the corporation's reluctance to afford

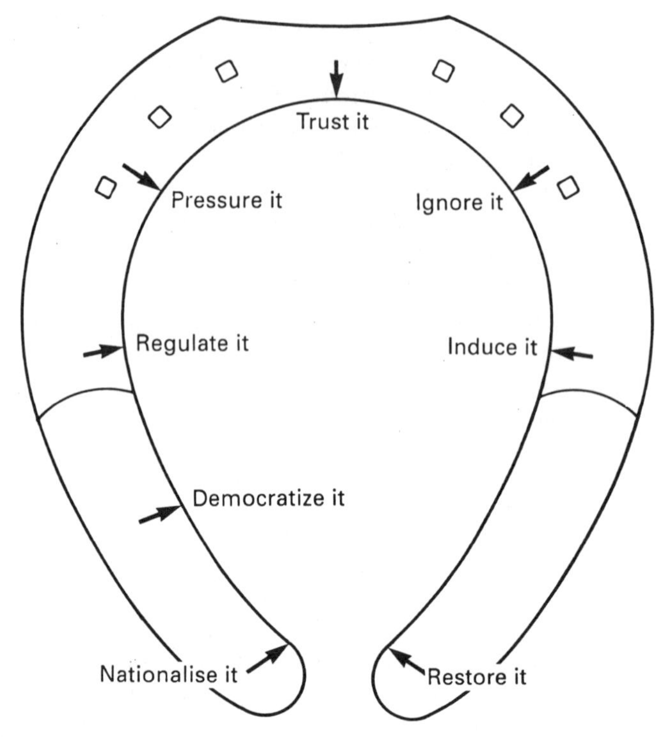

Figure 12.3 Mintzberg's conceptual horseshoe.
Source: Mintzberg (1989) p. 306.

stakeholders (including local publics) a powerful voice within the decision-making process. Unless there is a major shift in the base philosophy of business within the UK, the tendency will remain one of hiding behind the vexed issue of 'commercial confidentiality' and using technical expertise to support a predetermined policy position on environmental matters.

The imposition of *regulatory controls* also has serious deficiencies as an effective control strategy. Perhaps the greatest amongst these has been the apparent reluctance of the state to effectively underpin such legal developments with a powerful regulatory control infrastructure. One example of such 'failure' can be found in the hazardous waste industry. Here, attempts to impose tighter regulations in western countries have resulted in the export of such wastes to the developing world (see Smith and Blowers, 1991; 1992; Smith, 1991b). In the UK, attempts to tighten the regulatory framework have been beguiled by the lack of an effective policing strategy by an inspectorate that has been severely under-resourced in human terms until very recently, despite an ever-increasing stream of waste imports (see Smith, 1991a; Smith and Blowers, 1991; 1992). The dilemma here is that by increasing the regulatory controls in one country we might be encouraging industry to export its hazards elsewhere and that country may be less well equipped to deal with the problem.

The process of *pressurizing* the organization has long been used within democratic countries to exert some degree of control over organizations which appear to be acting outside the interests of society. The success of public opposition to technological and environmental problems has been variable, with corporations using their much greater power base to secure a favourable outcome (see, for example, Blowers, 1984; 1992; Crenson, 1974; Smith, 1990a). This imbalance of power is a key factor in undermining the likely validity of this particular control strategy. In terms of environmental problems the main drawbacks to this approach stem from a lack of publicly available information, the imbalance and legitimacy of technical expertise between public and corporate groups and the relative costs that can accrue to local groups during a lengthy public inquiry process.

Smith (1991b) has argued that as a consequence of the events during the 1980s, it would seem unwise to *trust* certain organizations to act in a socially responsible manner. The catalogue of disaster during that decade indicates that industry is often unable to control its own activities to the extent that it claims. This point is made quite forcibly by Tombs in his chapter on the chemicals industry (Chapter 10). The accidents at Bhopal, Chernobyl and Mexico City illustrated the fragility of corporate control mechanisms and, along with a series of financial scandals, have severely undermined the case for trust.

The severity of these events, including the insider and share-dealing incidents that occurred on both sides of the Atlantic, indicate that we cannot simply *ignore* the problem either. This is particularly the case with regard to many of the environmental problems that are currently on the political agenda as their impact is likely to be global, rather than simply local, and will affect future generations rather than simply our own.

Given the limitations of the control strategies detailed thus far it is obvious that the most effective way of regulating the corporation is to encourage, or as Mintzberg terms it, to *induce it* into behaving in a socially responsible manner. In terms of the disposal of hazardous waste, for example, there is considerable influence that can be brought to bear on the 'cowboy' waste companies by the more responsible chemical companies. This is particularly the case if 'cradle to grave' responsibility is introduced via legislative changes (Smith, 1991b; Smith and Blowers, 1991). Within this context it would become important for companies to conform, and be seen to conform, to the environmental demands of society.

The process of inducing the corporation to behave in a responsible manner is also present within the last of Mintzberg's control strategies, namely that of *restoring* the company to the control of the shareholders. Unfortunately such a move is likely to have little effect in terms of environmental control due to the range of shareholders and the concentration of voting power in small groups which may themselves be corporations or 'owner managers'. Again the waste disposal industry illustrates the problems associated with such an approach. Here the majority of companies are relatively small, with power firmly rooted in a small number of shareholders who also run the corporation on a daily basis. Similarly, the major energy utilities have until recently been in the hands of the state which has not shown itself to be accountable to the public (in this case the shareholders) over issues like nuclear power generation and waste disposal, acid rain and a whole range of hazard issues.

Given the problems associated with most of these strategies we need to address the question of which is the most effective in allowing us to maintain environmental quality and to move the corporation towards a more socially responsible position. Echoing the position adopted by Mintzberg (1983a; 1989) the answer has to be that no single approach is likely to prove successful. Mintzberg argues the case for a varied approach to the problem. He suggests that ultimately we have to have some trust in the organization and, whilst accepting the inherent limitations that exist within the concept of social responsibility, argues that,

> without honest and responsible people in important places, we are in deep trouble. We need to trust it because, no matter how much we rely on the other positions, managers will always retain a great deal of power. And that power necessarily has social no less than economic consequences.
>
> *(Mintzberg, 1983b, p. 111).*

Mintzberg continues his argument by suggesting that if such an approach fails then there is no alternative but to pressurize the firm and then regulate and induce it. Smith (1991b), in discussing the hazardous waste industry, has argued the case for a less 'liberal' approach to the problem. Here the suggestion is that we need to regulate the industry effectively and support that legislation with an efficient policing framework to ensure that companies comply. Given the recent history of environmental degradation in the developed countries it would seem that we can no longer completely trust the organization's belief in the power of its technocratic modes of decision-making (see Fischer, 1990) (whilst accepting that we may be able to trust individual managers therein) because it does not have the technocratic abilities that it lays claim to in areas of trans-science. Given the potential impacts that are associated with many of our industrial activities, we need to err on the side of caution and impose strict liability requirements on corporations at an *international* level rather than just within the boundaries of our own nation states. Indeed the importance of the left-hand side of the conceptual horseshoe is recognized by Mintzberg in his observation that

> a good deal of what passes for social responsibility would disappear without other, countervailing forces on the corporation – pressure campaigns by activists, regulations by the government, and so on. Much so-called enlightened self-interest would become far less enlightened if the likes of Ralph Nader did not lurk outside the gates of every large corporation.
>
> *(Mintzberg, 1983, p. 12)*

The Greening of Business – Beyond Rhetoric?

This final chapter has sought to argue the case for a more socially responsive corporate strategy in the belief that such an approach lies at the centre of the current debates surrounding business and the environment. Whilst there is little doubt that the scale of the problem is of increasing concern to society a major factor in preventing rapid action by industry concerns the nature and extent of the process of corporate responsibility.

The present political climate requires that organizations seek to develop a more responsive approach to the demands of the public. Without such an approach the response from industry will be constrained by a legal and economic definition of its legitimacy as defined by those who argued the case for corporate obligation rather than responsibility. Within this debate there is considerable scope for business educators to take a leading role in changing and shaping the dominant managerial paradigm that operates within a restrictive view of corporate legitimacy.

An important factor within this process is the values and attitudes held by the current generation of managers, many of whom have been educated within an environment that has been overly influenced by the ideas of Friedman. In order to achieve a more socially responsive generation of managers we need to address the fundamental question of how they are educated. This point is already being addressed by a number of the more proactive business schools, in both the USA and UK, who are seeking to change their culture rather than simply 'bolt on' environmental issues to existing courses (see Smith, 1990c; Smith and Hart, 1991). To date, however, the take-up of such ideas has been slow and will need to show a substantial increase if we are to achieve more than a temporary move towards corporate responsiveness. There is no doubt that in order to change the deep core values of business we will have to change the values of those who will manage them throughout the 1990s and beyond. Similarly, we need to equip business educators with the skills and expertise necessary to change the core of their academic courses to include concepts of corporate responsibility and responsiveness. Such a fundamental shift will be an important prerequisite if we are to prevent the issues from simply being passing phenomena. A socially responsive corporate culture would seem to be the only way out of the current wave of environmental problems.

This book marks an attempt to raise issues for the agenda – for both corporate mangers and business educators alike – rather than provide simple solutions to complex problems. The hope is that corporations will, in time, become more responsive to the environmental issues that face them and that we can break out of the cycle of rhetoric and learn to manage business within a 'greener' corporate culture.

References

Ackerman, R. W. and Bauer, R. A. (1976) *Corporate Social Responsiveness*, Reston Publishing, Reston, Virginia.

Allen, J. F. (1907) *Some Founders of the Chemical Industry*, cited in K. Warren (1980) *Chemical Foundations: The Alkali Industry in Britain to 1926*, Oxford University Press.

Blowers, A. (1984) *Something in the Air: Corporate Power and the Environment*, Harper & Row, London.

Blowers, A. (1992) Narrowing the options: the political geography of waste disposal, in M. Clark, D. Smith and A. Blowers (eds.) (1992) *Waste Location: Spatial Aspects of Waste Management, Hazards and Disposal*, Routledge, London, pp. 227–47.

Carroll, A. B. (1979) A three-dimensional conceptual model of corporate performance, *Academy of Management Review*, Vol. 4, no. 4, pp. 497–505.

Crenson, M. A. (1974) *The Unpolitics of Air Pollution: Decision-Making in the Cities*, Johns Hopkins Press, Baltimore, MD.

Davis, K. (1975) Five propositions for social responsibility, *Business Horizons*, June, pp. 19–24.

Epstein, E. E. (1987) The corporate social policy process: beyond business ethics, corporate social responsibility, and corporate social responsiveness, *California Management Review*, Vol. 29, no. 3, pp. 99–114.

Fischer, F. (1990) *Technocracy and the Politics of Expertise*, Sage, Newbury Park, CA.

Fisher Unwin, T. (1888) *Industrial Rivers of the United Kingdom*, cited in K. Warren (1980) *Chemical Foundations: The Alkali Industry in Britain to 1926*, Oxford University Press.

Freeman, R. E. and Gilbert, D. R. (1988) *Corporate Strategy and the Search for Ethics*, Prentice Hall, London.

Freeman, R. E., Gilbert, D. R. and Hartman, E. (1988) Values and the foundations of strategic management, *Journal of Business Ethics*, Vol. 7, pp. 821–34.

Friedman, M. (1962) *Capitalism and Freedom*, University of Chicago Press.

De George, R. T. (1986) Theological ethics and business ethics, *Journal of Business Ethics*, Vol. 5, pp. 421–32.

Gilbert, D. R. (1986) Corporate strategy and ethics, *Journal of Business Ethics*, Vol. 5, pp. 137–50.

Goodpaster, K. E. (1983) The concept of corporate responsibility, *Journal of Business Ethics*, Vol. 2, pp. 1–22.

Hajer, M. A. (1990) The discursive paradox of the new environmentalism, *Industrial Crisis Quarterly*, Vol. 4, no. 4, pp. 307–10.

Jones, T. M. (1980) Corporate social responsibility revisited, redefined, *California Management Review*, Vol. 22, no. 2, pp. 59–66.

McGrew, A. (1990) The political dynamics of the new environmentalism, *Industrial Crisis Quarterly*, Vol. 4, no. 4, pp. 291–305.

Mintzberg, H. (1983a) The case for corporate responsibility, *Journal of Business Strategy*, Vol. 4, no. 2, pp. 3–15.

Mintzberg, H. (1983b) Why America needs but cannot have corporate democracy, *Organizational Dynamics*, Spring 1984.

Mintzberg, H. (1989) *Mintzberg on Management: Inside our Strange World of Organizations*, Free Press, New York.

Murray, K. B. and Montanari, J. R. (1986) Strategic management of the socially responsible firm: integrating management and marketing theory, *Academy of Management Review*, Vol. 11, no. 4, pp. 815–27.

Sethi, S. P. (1975) Dimensions of corporate social performance: an analytical framework, *California Management Review*, Vol. 17, no. 3, pp. 58–64.

Shelley, M. (1818) *Frankenstein*, Penguin, Harmondsworth.

Shrivastava, P. (1992) *Bhopal – The Anatomy of a Crisis*, 2nd ed., Paul Chapman, London.

Smith, D. (1990a) Corporate power and the politics of uncertainty: conflicts surrounding major hazard plants at Canvey Island, *Industrial Crisis Quarterly*, Vol. 4, no. 1, pp. 1–26.

Smith, D. (1990b) Beyond contingency planning: towards a model of crisis management, *Industrial Crisis Quarterly*, Vol. 4, no. 4, pp. 263–75.

Smith, D. (1990c) Green into gold, *The Times Higher Education Supplement*, 15 June p. 14.

Smith, D. (1991a) Beyond the boundary fence: decision-making and chemical hazards, in

J. Blunden and A. Reddish (eds.) (1991) *Energy, Resources and the Environment*, Hodder and Stoughton, London, pp. 267–91.

Smith, D. (1991b) The Kraken wakes – the political dynamics of the hazardous waste issue, *Industrial Crisis Quarterly*, Vol. 5, no. 3, pp. 189–207.

Smith, D. and Blowers, A. (1991) Passing the buck – hazardous waste as an international problem, *Talking Politics*, Vol. 4, no. 1, pp. 44–9.

Smith, D. and Blowers, A. (1992) Here today, there tomorrow: the politics of hazardous waste transport and disposal, in A. Clark, D. Smith and A. Blowers (eds.) (1992) *Waste Location: Spatial Aspects of Waste Management, Hazards and Disposal*, Routledge, London, pp. 208–26.

Smith, D. and Hart, D. (1991) The greening of business education in the UK. Paper presented at the 11th Annual International Conference of the Strategic Management Society, 'The Greening of Strategy-sustaining performance', Toronto, October 23–26.

Walters, K. D. (1977) Corporate social responsibility and political ideology, *California Management Review*, Vol. 19, pp. 40–51.

Wilson, I. (1974) What one company is doing about today's demands on business, in G. A. Steiner (ed.) (1974) *Changing Business-Society Interrelationships*, Graduate School of Management, UCLA, Los Angeles.

INDEX